Chemistry of

Robert H. Doremus
Gary E. Wnek
Mark Palmer
Rensselaer Polytechnic Institute

McGraw-Hill, Inc.
College Custom Series

*New York   St. Louis   San Francisco   Auckland   Bogotá
Caracas   Lisbon   London   Madrid   Mexico   Milan   Montreal
New Delhi   Paris   San Juan   Singapore   Sydney   Tokyo   Toronto*

CHEMISTRY OF MATERIALS

1 2 3 4 5 6 7 8 9 0 HAM HAM 9 0 9 8 7 6 5 4

ISBN 0-07-017975-1

Editor: Gale Murray

Cover Design: Christopher Siwinski

Printer/Binder: HAMCO Corporation

CONTENTS

CONTENTS

1

INTRODUCTION

1.1 ➤ INTRODUCTION

There has never been a more exciting time to study chemistry. Advances in microelectronics, medical diagnostics and high-temperature superconductors have been made possible by the ability to create new materials with new (and sometimes unexpected) properties. Macroscopic properties such as electrical conductivity, stress, fracture resistance, and optical transparency are important to engineers for the design of new products and systems. It is important to recognize that each of these properties has an atomic/molecular level origin. Therefore, in order to really exploit the limits of what materials can do, and design materials with new properties, it is necessary to have an appreciation of phenomena which occur on the microscopic level of atoms and molecules. This brings us into the domain of <u>chemistry</u>. Our aim in this two-semester course is to integrate the microscopic view of matter with macroscopic properties of interest to all engineers.

A crude definition of a 'material' is a substance that has utility in the solid state. As summarized in Figure 1, the properties and ultimately the performance of a material are intimately linked to atomic level structure and atomic composition. Structure and composition depend on the synthesis pathway chosen and how the material is processed or fabricated.

The first semester of our course, with its emphasis on bonding, thermodynamics, kinetics, and an introduction to solids, prepares you for our focus on the chemistry and properties of materials. The second semester begins with 'real' (defect-containing) solids, and then turns again to thermodynamics and kinetics but with a focus on solids. We will then spend a fair bit of time on classes of materials and their properties, and end with several examples of materials systems (e.g., the semiconductor chip), where several different types of materials are employed to achieve a particular application.

Throughout the course, we will introduce many practical examples justifying why what we teach you is relevant and important.

Our course outline was developed in a sequence that we believe makes a lot of sense. However, this course is unique, and therefore there is no single textbook which adequately covers the subject matter. The chapters to follow represent an our efforts to highlight the important ideas. We have also selected a very readable textbook (Chang, "Chemistry," 5th Edition, McGraw-Hill, 1994) to complement our chapters. The important sections to read can be found in the detailed course outline.

1.2 ➤ CLASSES OF MATERIALS

Materials are frequently classified into five broad categories: metals, ceramics, polymers, electronic materials (primarily semiconductors), and composites (multiphase materials). Characteristic properties of each are summarized below.

Metals
- High electrical and thermal conductivities.
- Wide range of mechanical properties.
- Lustrous.
- Properties can be tailored through alloying and/or processing.

Ceramics
- Usually low electrical and thermal conductivities.
- Excellent thermal stability.
- Typically brittle, but can be toughened.
- Properties can be tailored through composition control and/or processing.

Polymers
- Wide range of mechanical properties (brittle to rubbery).
- Light weight an advantage.
- Properties can be tailored through chain composition, sequence distribution, isomerism, and/or processing.

Electronic Materials
- Conductivities can be controlled through 'doping.'
- Typically brittle.
- Circuits can be patterned using chemistry.

Composites

- Combine properties of a matrix material and a reinforcing material. Properties can be superior to those of individual components.
- Broad control of properties depending on nature and proportions of components.

Problem 1-1

Provide two examples of each class of material on the previous page. Discuss an application of each example and why that material was selected for that application.

1.3 ➤ EXAMPLES OF MATERIALS AND THEIR APPLICATIONS

Engineers who build engines have known for many years that fuel efficiency increases the hotter an engine runs. High operating temperatures, however, demand materials that can last without melting or suffering from thermal decomposition. Ceramics usually offer superior thermal stability compared with other materials although they are notoriously brittle, making them undesirable for engine components. Two choices exist regarding ceramics. First, we could simply accept that ceramics are brittle and limit their use to applications where mechanical properties are not critical. The second choice is to understand at the atomic level (e.g., chemistry) the reasons for brittleness and then create clever solutions to mitigate this problem. Indeed, this is what engineers have done, and more and more we are seeing ceramics being used in situations which we not imagined a decade ago. Incidentally, Mother Nature is frequently well ahead of us in problem solving. For example, clam shells are about 99% by weight $CaCO_3$ (limestone, a ceramic), and yet they are very tough, especially when wet. Limestone is a brittle material, and yet the plates of limestone glued together with protein in the clam shell leads to a very tough material.

Polymers, of which plastics represent a major component, are pervasive because of their low cost, light weight and ease of fabrication into many different shapes. Compact disks are possible because polymers are available that are optically clear and relatively tough. The polymer Kevlar has extremely high strength and extraordinarily good thermal stability for an organic-based material and is used as a reinforcing agent in composites. Advantage is taken of the temperature resistance of Kevlar in its use as a fabric for firefighters' clothing.

The electronics industry, and society as a whole, has witnessed a revolution has led to powerful computers and calculators as well as CD players and telecommunications equipment, all made possible by the semiconductor chip. This is basically a piece of silicon in which small regions have been manipulated chemically to create devices such as diodes and transistors. The silicon chip (an example of an electronic material) is first patterned with the help of thin polymer coating which is light-sensitive. A 'window' is cut through the layer of SiO_2 (a ceramic) that is naturally present on the silicone surface, and small amounts of impurity atoms are diffused in to form the electronic devices. Finally, metal contacts are made to connect the chip to the outside world. Notice that in this example materials from four different classes had to be considered.

These and other examples will be discussed in detail in our course. Stay tuned.

1.4 ➤ GETTING STARTED—A BRIEF SUMMARY OF SOME BASIC IDEAS OF CHEMISTRY

(1) Le Chatelier's Principle. This essentially states that if a system at equilibrium is disturbed in some way, it will try to adjust itself to relieve the disturbance. We'll see many examples of this idea in the course.

(2) Conservation of mass. Mass can neither be created nor destroyed.

(3) Conservation of energy. energy can neither be created nor destroyed but only changed from one form to another (e.g., electrical to thermal).

(4) The mole and stoichiometry. You are used to counting, say, eggs or doughnuts by the dozen. The number twelve is assigned to this term. What about the counting of atoms or molecules? We count these in moles, and one mole contains 6.02×10^{23} atoms or molecules. This is a huge number, which is not too surprising given that atoms are very small. Where does this number, referred to as Avogadro's number, come from?

By international agreement, the mass of a single carbon-12 atom is defined as 12 u (u is a 'unified atomic mass unit'). It turns out that 6.02×10^{23} u has a mass of 1 g. Suppose you had 12 g of carbon-12. Since each atom would weigh 12 u, there will be 6.02×10^{23} atoms of carbon-12 in this sample. The twelve grams of carbon-12 represent <u>one mole</u> of that material. The mole is the amount of any substance that contains the same number of elementary particles as there are atoms in exactly 12 g of carbon-12.

The best way to get the idea is to do a few problems:

Problem 1-2

Write a balanced reaction for the production of water from diatomic hydrogen and diatomic oxygen. How many moles of hydrogen are needed to make one mole of water? How many moles of oxygen are needed? What is the relationship between the number of molecules reacting and the corresponding number of moles reacting?

(Note: The halogens (look up the term), oxygen, nitrogen and hydrogen are important gases. You should realize that they consist of diatomic molecules. The inert or noble gases, on the other hand, are monatomic.)

Problem 1-3

Write a balanced reaction for the combustion of methane to carbon dioxide and water. How many moles of water and CO_2 are produced per mole of methane burned?

Problem 1-4

Your roommate claims to have three moles of paper clips in his/her desk. How many paper clips would this amount to?

Problem 1-5

Recall that oxidation involves the loss of electrons and reduction the gain of electrons. How many moles of electrons are required to reduce one mole of chlorine gas to two moles of chloride ions?

Problem 1-6

Given the following reaction:

$$2\,UF_5 \;+\; 2H_2O \;\rightarrow\; UO_2F_2 \;+\; UF_4 \;+\; 4HF$$

Calculate the mass of UF_4 formed from 10 g of UF_5.

(5) Oxidation Numbers. Elements usually have a tendency to lose or gain electrons. For example, sodium will easily lose one electron to form sodium ion, Na^+. (Sodium is easily oxidized, and therefore is a good reducing agent.) The oxidation number of Na^+ is +1. Chlorine readily gains one electron (and therefore is a good oxidizing agent), and the Cl^- ion is said to have an oxidation number of -1. Your periodic table will help in identifying stable oxidation numbers for various elements. (By definition, the oxidation number of a pure element is zero.)

More problems:

Problem 1-7

Assign oxidation numbers to the atoms in the following: Al_2O_3 (aluminum oxide), Al metal, water, phosphorous trichloride, phosphorous pentachloride, lithium hydride (LiH).

Problem 1-8

Write a balanced chemical reaction for the oxidation of iron metal to FeO. Note the oxidation numbers of atoms in the product and reactants. Can the FeO be oxidized further? If so, write a balanced reaction.

Problem 1-9

a) Write a balanced chemical reaction for the oxidation of aluminum metal to aluminum oxide. Note the oxidation numbers of atoms in the product and reactants. Can the Al_2O_3 be oxidized further? If so, write a balanced reaction.

b) Calculate the mass% of Al and O in Al_2O_3.

c) If you had one mole of oxygen gas and one mole of aluminum metal, how many moles of Al_2O_3 could you make? Which is the limiting reactant, oxygen or aluminum?

Brain Teaser: We're used to thinking that atoms exist in compounds in ratios of whole numbers. However, there are many materials which apparently violate this rule. For example, $Fe_{0.95}O$ exists. How is this composition possible? (Hint: a transition metal like iron can exist in more than one oxidation state.)

2

ATOMIC STRUCTURE

2.1 ➤ HISTORY

Towards the end of the nineteenth century there was considerable experimental evidence for atoms as the smallest unchanged division of matter. There were still many scientists, however, who did not accept these ideas. One of the papers written by Albert Einstein in 1905, his astoundingly productive year, was designed to show the atomic nature of the matter. This paper on Brownian motion, the random motion observed in a microscope of small particles in a liquid, showed that this motion resulted from collisions of unseen molecules of the liquid with the particles. Today this paper is still basic to understanding of motion in a liquid, but Einstein's main purpose in writing it was to convince doubters that atoms and molecules really did exist.

2.2 ➤ ATOMIC NUCLEUS AND ELECTRONS

At the same time that the idea of atoms and molecules as the ultimate units of matter was being accepted, other experiments were showing that atoms had a complex finer structure on a scale much below the diameter of an atom of about .3nm. This substructure of the atom is composed of three different particles, protons, neutrons, and electrons. The protons and neutrons are bound tightly in a tiny nucleus at the center of the atom, and the electrons are around the nucleus. The mass of the atom is almost entirely in the nucleus; the mass of an electron is about 1/1840 that of a proton or neutron. Nevertheless the diameter of the nucleus is less than one ten thousandth of the diameter of the atom. If the nucleus were the size of a pea, the atom would be as large as a football field. The protons are positively charged, the neutrons are uncharged, and the electrons are negative. Since the atoms are neutral, the number of protons and electrons is the same and the number of neutrons is variable.

The force that holds the electrons and nucleus together is the electrostatic ("coulombic") attraction between the positively charged nucleus and the negatively charged electrons around it. Highly precise experiments have shown that the charge on an electron is exactly equal and opposite to the charge on the proton.

It is very surprising that the positively charged protons bind together tightly in the atomic nucleus. From consideration of electrostatic forces alone, one would expect a strong repulsion between protons. Thus, there must be another kind of force holding the nucleus of neutrons and protons together. Gravitational forces are far too weak to hold the nucleus together. The nuclear binding force is strongly attractive at the short range (about 10^{-13}cm). At even smaller separation $(.5(10)^{-13}$cm) the interaction between nucleons is repulsive, so each proton or neutron occupies a volume that can be approximated by a sphere 10^{-13}cm in diameter.

The atomic number Z of an atom is the number of protons in its nucleus. For example, copper atoms have 29 protons and 29 electrons. The mass number of an atom is the total number of protons plus neutrons in the atomic nucleus. All atoms except hydrogen contain neutrons as well as protons. The nucleus of normal hydrogen contains one proton and no neutrons. Heavy hydrogen has one proton and one neutron in its nucleus. Normal and heavy hydrogen are called isotopes of hydrogen. A heavier atom such as copper can have several isotopes. Thus, copper can have 34 or 36 neutrons in its nucleus in addition to the 29 protons. Natural copper is a mixture of 69.1% copper 63 (which has a nucleus with 29 protons plus 34 neutrons, equaling 63 particles) and 30.9% copper 65. Another example is silicon, with three stable isotopes, silicon 28, 29, and 30, containing 14 protons and 14, 15, and 16 neutrons, respectively.

2.3 ➤ ATOMIC MASS

The mass of an atom is close to the sum of the masses of the protons and neutrons in its nucleus. The electrons contribute a very small additional mass, which is negligible for most purposes. The international standard of mass is the isotope carbon-12, which is defined as a mass of 12 atomic mass units (amu). The masses of individual isotopes in amu are close to the sum of the number of protons and neutrons but not exactly. Thus, copper 63 has a mass of 62.93 amu and copper 65 a mass of 64.93 amu.

The mass of an atomic nucleus is always slightly smaller than the sum of the masses of the protons and neutrons in the atom when they are isolated. This mass difference comes from mass that was changed into energy upon formation of the nucleus from the neutrons and protons; this energy is released from the atom. The same amount of energy is needed to break apart the atom into individual neutrons and protons, and is called the binding energy of the nucleus. The relationship between the binding energy E and the mass difference 63 and 62.93 can be calculated from the Einstein relation between mass and energy:

$$E = mc^2 \tag{1}$$

in which E is the energy equivalent to mass m, and c is the speed of light $(3(10)^8 \text{m/s})$.

Sample Problem:

Calculate the binding energy of copper 63 from the masses of neutrons and protons in the nucleus. The mass of a proton is 1.0073 amu and a neutron is 1.0087 amu.

The total mass of the separated protons and neutrons is:

$$
\begin{aligned}
1.0073 \times 29 &= 29.2117 \quad \text{protons} \\
1.0057 \times 34 &= \underline{34.2958} \quad \text{neutrons} \\
& \ 63.5075
\end{aligned}
$$

one amu = $1.66(10)^{-27}$kg.

The difference between the mass of the actual copper 63 nucleus 62.93 and 63.51, the sum of masses of the isolated nucleons is .58 amu. Then from Eq. 1

$$
\begin{aligned}
E &= (.58)(1.66)(10)^{-27} \times 9(10)^{16} \\
&= 8.67(10)^{-11} \text{ joules}
\end{aligned}
$$

The average atomic mass or atomic weight of a natural material depends upon the percentage of different isotopes it contains. Thus, natural copper contains 69.09% of copper 63 and 30.91% of copper 65, so its average atomic mass is .6909 × 62.93 + .3091 × 64.93 = 63.55 amu. This number of 63.55 corresponds to the atomic weight of copper found in tables of atomic weight.

2.4 ➤ ATOMIC SPECTRA

One of the best ways to study the electronic energy states of an atom of an element is to measure the spectrum of light emitted or absorbed by atoms of the element in the vapor. To measure the emission spectrum of an atom it must be first vaporized in sufficient quantity to be observed. Although few atoms of an element such as a metal vaporize at room temperature, they do vaporize at higher temperatures. The spectra of atoms in gases such as the inert gases helium, neon and argon can be studied even at room temperature.

The experimental arrangement for measuring emission spectra is shown in Fig. 7.6, p. 255, Chang. Vaporized atoms are heated to a temperature high enough for them to emit light, and the wave length (color) of the emitted light is studied. The emitted light can be broken into its wave length components with a prism, and the intensity of emission at different wave lengths measured with a light detector or photographic plate. Often the emission is stimulated by passing electrons through the gas at lower temperature, as shown in the figure. When an electron collides with an atom, the atom absorbs energy from the electron.

The absorption of light by a vapor of atoms can also be studied. In this experiment light beams of different wave lengths are passed through the vapor, and the fractions of the beams that are absorbed by vapor atoms are measured with a detector or photographic plate.

Atomic emission is useful in a number of practical applications. The yellow light of street lamps comes from emission from a vapor of heated sodium atoms, and blue street lamps have emitting mercury atoms. The familiar red neon lamps are colored by emission from neon atoms.

An emission spectrum of hydrogen is shown in Fig. 7.8, p. 256, Chang. From traditional physics one would expect a continuous band of light being emitted from the hydrogen atoms. The result, however, is entirely different. A series of sharp lines of different wave length (color) are emitted, and many regions are dark. This experimental result shows that the internal components of an atom behave in a way quite different from what is expected from the behavior of particles of material or even of molecules in a gas. Experiments show that molecules in a gas can have any energy in the range below the highest energy. The measurements of atomic spectra suggest that the electrons in an atom behave differently from molecules in a gas.

2.5 ➤ PHOTOELECTRIC EFFECT

The results of another kind of experiment involving the electrons in atoms also are different from what was expected from classical physics. When light is shined on a solid surface, such as clean metal, electrons are ejected from the surface and can be collected on a positively charged target, as shown in Fig. 7.5, p. 253, Chang. If the electrons followed classical mechanics, one would expect that their energies would increase as the intensity of the light increased. The experimental result, however, is that the energies of the electrons do not depend on the intensity of the light, but only on the wave length of the light.

2.6 ➤ WAVES IN A STRING OR WIRE

The two above experiments and others showed that the behavior of electrons in atoms is quite different from what is expected from traditional mechanics. The explanation of these effects came from the development of one of the monumental achievements of the human mind, quantum mechanics. To reach some understanding of what quantum mechanics tells us about the behavior of electrons, we must learn something about waves, because electrons are found to have characteristics of both particles and waves.

Consider a string held at its ends by two different persons. If one person moves the string up and down, a wave moves along the string. This motion is called a traveling wave. On the other hand, if the string is pulled taut and then plucked in the middle, it vibrates, as shown in Fig. 7.16, p. 262, Chang. The same thing happens when a piano wire or violin string is struck or plucked. It vibrates only with particular frequencies, giving a particular tone. In this vibration the wave length is an integer multiple of L/2, where L is the distance between the ends of the string, as shown in the figure. Short wave lengths that are integral multiples of L/2 are called harmonics, and give higher tones. The connection with atomic spectra comes in the observation that the frequency of vibration (energy) of the wire is fixed by the properties of the wire and the "boundary conditions" (tension and length of the wire). Other frequencies are excluded, just as certain light frequencies are not present in atomic spectra.

2.7 ➤ ELECTROMAGNETIC WAVES

Radio waves are vibrations of an electrical field (potential gradient) through space. The form of these waves is similar to the waves in a string, except that they propagate over long distances. The Scottish physicist, James Clerk Maxwell showed that electrical and light waves both result from the vibration of electrical fields in space. The only difference between them is their frequency or wave length of vibration. The spectrum of electromagnetic vibrations is shown in Fig. 7.4, p. 252, Chang. It includes waves of very high frequency, from gamma rays, through x-rays, light rays, and radio waves to low frequency electrical waves.

The frequencies v and wave lengths λ of all these electromagnetic waves are related by the equation:

$$v = c/\lambda \qquad (2)$$

in which c is the velocity of propagation of the wave. In a vacuum, and very nearly in air, this velocity is $3.00(10)^{10}$cm/s or $3.00(10)^{8}$m/s.

In hindsight we can see that the lines in atomic spectra and the particular frequencies of a vibrating wire suggest that electrons could have the character of a wave. However, physicists resisted this connection for many years because it is so contrary to our intuition and experience to associate wave behavior with a particle such as an electron. Finally in 1924, Louis de Broglie, who was a French physicist and aristocrat, postulated that electrons have associated with them a wave length λ. If the electron has a mass m and a velocity v, the wave length λ is inversely proportional to the product mv:

$$\lambda = h/mv \qquad (3)$$

in which h is a universal constant called Planck's constant. In classical mechanics of particle motion the product mv is called the momentum, so de Broglie's postulate can also be stated that the momentum of an electron is equal to h/λ.

The result of de Broglie's idea is that an electron has characteristics of both a particle (it has mass) and of a wave. This dual character of an electron has profound influence on its behavior and properties, such as energy and distribution in space. The consequences of this duality were first worked out in two different ways by two German physicists Werner Heisenberg and Erwin Schrödinger. The result of this new mathematical treatment of the behavior of electrons was called quantum mechanics or wave mechanics. We will consider certain important conclusions from this treatment, especially about the energies and spacial distribution of electrons, without going into the details of the mathematical treatments that lead to these conclusions. Thus, the reader must accept these conclusions without proof; you will find that they are essential to understanding a large variety of physical and chemical phenomena. Examples are the chemical behavior of atoms, bonding between atoms into molecules, and the atomic structure of molecules, liquids and solids.

2.8 ➤ ENERGIES OF ELECTRONS

Quantum mechanics shows that electrons in atoms have only particular energies; in other words the energies are quantitized. The hydrogen atom is not the normal state of hydrogen, which usually has two atoms in a molecule, but is the simplest atom because it has just one proton in the nucleus and one electron. Thus, the mathematical equations for the behavior of the electron in the hydrogen atom can be solved rigorously, so we will use this atom as an example of the results of the quantum mechanical behavior of electrons.

The electron in the hydrogen atom normally has an energy

$$E = \frac{me^4}{8\epsilon_o h} \tag{4}$$

In this equation m and e are the mass and charge of the electron, ϵ_o is the dielectric constant of a vacuum, and h is Planck's constant. For the numerical values of these parameters and their units see the last page in Chang. This energy has a value $2.16(10)^{-18}$ joule, or 13.5 electron-volts; it is also sometimes made a unit itself called one rydberg. This energy is called the ground state energy of the electron in the hydrogen atom. If the electron receives additional energy from an external source, for example from a light wave, it becomes excited into a state of higher (more positive) energy called an <u>excited state</u>. Quantum mechanics shows that these energies of excited states can have only particular values determined by integers n called quantum numbers. These energies are

$$E_n = -\frac{me^4}{8\epsilon_o h}\left(\frac{1}{n^2}\right) \tag{5}$$

In other words these energies are just the ground state energy (quantum number n=1) divided by the squares of integers, for example 4, 9, 16, 25, etc. These energies are shown in levels in Fig. 7.11, p. 258, Chang.

These quantitized energies of the electron give an explanation of the lines in the emission spectrum of an atom. For example, there is a strong line in the emission spectrum of hydrogen atoms at a wave length of .1215 μm, which corresponds to an energy of 10.2 eV. This line results from transitions of electrons from the first excited state (n=2) to the ground state (n=1); the change of energy is $\Delta E = 13.5(1/1 - 1/4) = 10.2$ eV. Other lines in the spectrum result from other transitions between different energy levels of electrons in the atom.

2.9 ➤ ELECTRON IN A BOX

One of the simplest problems to solve for the behavior of an electron is the electron in a constrained space with a constant electrical potential, which is called an <u>electron in a box</u>.

The waves that go with the electronic this situation are just like the standing waves in a string or wire that is fixed at its ends. This problem may not seem to have much value, but it is actually a very useful model for the electrons in metal. Many properties of the electrons in a metal can be understood from the three-dimensional version of this electron-in-a-box problem. The energies of the electron in a box are:

$$E = \frac{h^2}{8mL^2} n^2 \tag{6}$$

in which L is the length of the box and n is the quantum number. The ground state (n=1) is $3.8(10)^{-15}/L^2$ev, with L in centimeters. For L equal one centimeter the ground-state energy is nearly zero, but for L equal .3nm (atomic dimensions) it is about 4 ev, which is a substantial energy.

2.10 ➤ DISTRIBUTION OF AN ELECTRON AROUND AN ATOM

The distribution of an electron around an atom as deduced from quantum mechanics is quite different from the distribution expected from classical mechanics. One of the first ideas to explain the properties of atoms was that the electrons circled the nucleus like planets around the sun (Bohr atom). However, the electrostatic attraction between the electron and the nucleus is so strong that the electron should circle into the nucleus in a short time in this model. The quantum mechanical distribution is described by the probability of an electron being at a particular distance. Let us first consider the spherically symmetrical distribution of the electron in the ground state (n=1) of the hydrogen atom. The distribution can be thought of as a "cloud" around the nucleus. If we draw a line out from the nucleus in any direction, the probability of finding an electron at a particular distance is shown in Fig. 2-1 (see page 16). There is a maximum probability at a particular distance, which is about .053 nm for the electron in the hydrogen atom. This distance is often taken as the "radius" of the hydrogen atom thought of as a sphere, but Fig. 1 shows that this is a considerable simplification. The atom is actually "squishy"; the electron has a considerable probability of being within a cloud near the nucleus. The electron cannot be at the nucleus, just where the electrostatic attraction should place it, and it has a finite probability (although very small) of being at any distance from the nucleus.

The electron cannot be placed at a particular distance from the nucleus at a particular time; all that can be deduced is the probability of the electron being at a particular position. This lack of certainty or "causality" has generated a great deal of philosophical agonizing about quantum mechanics and even some resistance to this idea of a probability distribution. Albert Einstein spoke for this concern when he said, "God does not play dice." Nevertheless this idea of the position of the electron is an integral part of quantum mechanics and has been widely accepted.

For quantum numbers n (positive integers) greater than one there are a fascinating variety of electron distributions possible. Although the energy of the electron in the hydrogen atom is uniquely determined by the quantum number n and Eq. 5, there are additional quantum numbers that lead to different probability distributions of the electron around the atom. Quantum numbers l correspond to a different distribution of electron probabilities around the nucleus; these different distributions are called orbitals, and are designated by letters. This correspondence of l values and orbital letters is shown in Table 2-1. The names arose from the lines in atomic spectra that correspond with transitions involving the l quantum numbers.

2.11 ➤ RELATION WITH THE PERIODIC TABLE

The discussion of electronic properties has centered so far on the hydrogen atom only. This atom is scientifically interesting, but not of great practical importance. One might expect that extension of the quantum mechanical properties to larger atoms with more electrons would be difficult. A highly striking result is that the properties of the electrons in the various excited states of the hydrogen atom can be generalized to electrons in atoms with many electrons. The electrostatic repulsion of these electrons contributes only a small perturbation to their behavior, so that the electron states for the hydrogen atom can be used for large atoms as a first approximation. The way that this is done to generate the periodic table is described in the next chapter. The remarkable result is that the periodic table can be derived from the hydrogen quantum numbers, giving a profound understanding of the chemistry of different elements and of atomic bonding.

The directional properties of the quantum mechanical distribution functions have a striking correspondence with bonding between atoms. A spherically symmetrical distribution of electrons about the nucleus corresponds to bonding between ions, which is electrostatic and behaves like the attraction and repulsion between charged spheres. The bonding in a solid metal, although more complicated, also behaves like interaction between spheres. The specific bonding between atoms by sharing of electrons, called covalent bonding, takes place along particular directions. An example of covalent bonding is in the methane CH_4 molecule. Each hydrogen atom is bonded to the central carbon atom, and the bonds are equally spaced apart, as shown in the figure at the top of p. 377, Chang. An example of covalent bonding in a solid is alumina, Al_2O_3. Each aluminum atom is bonded to six oxygen atoms; the bonding direction corresponds with the electron distribution of p orbitals shown in Fig. 7.25, p. 270, Chang. Thus, quantum mechanics and the structure of atoms determines the chemical behavior of atoms and molecules and the structure of solid materials.

FIGURE 2-1

Probability of finding an electron at a particular distance from the nucleus of a hydrogen atom, for the lowest energy state (n=1).

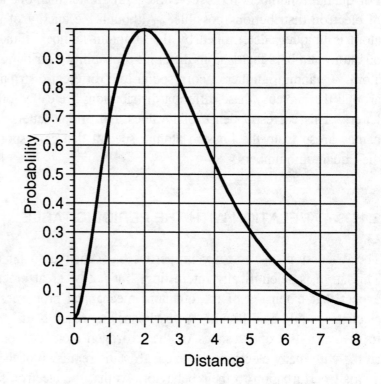

TABLE 2-1
Correspondence of Values of the l Quantum Number and Orbital Letters

l	orbital name	name from atomic spectroscopy
0	s	sharp
1	p	principle
2	d	diffuse
3	f	fundamental
4	g	
5	h	

PROBLEM SET #1 ➤ *Nuclear Structure*

Solved Problem 1: Determine the number of protons, neutrons, and electrons in the following: ^{24}Mg, ^{56}Fe, $^{24}Mg^{2+}$, and $^{56}Fe^{2+}$.

The number of protons is equal to the atomic number which is found on the periodic table. Mg has an atomic number of 12, and thus has 12 protons. Fe has an atomic number of 26, and 26 protons, Mg^{2+} is still magnesium has an atomic number of 12, and therefore 12 protons. Similarly Fe^{2+} has 26 protons.

The atomic mass number is the superscripted prefix, the atomic mass number is the number of nucleons (protons and neutrons) that are present in the nucleus. ^{24}Mg has 24 total nucleons in the nucleus 12 of these are protons as shown on the periodic table. Therefore, ^{24}Mg has (24-12) or 12 neutrons. ^{56}Fe has an atomic mass number of 56, indicating 56 protons or neutrons in the nucleus. 26 of these are protons as shown on the periodic table therefore, ^{56}Fe has 30 neutrons. The atomic mass number for $^{24}Mg^{2+}$ is the same as that of , ^{24}Mg so $^{24}Mg^{2+}$ has 12 neutrons. Similarly $^{56}Fe^{2+}$ has 30 neutrons.

In an atom there is no net electrical charge, it is neutral. Therefore there must be an equal number of protons and electrons. Mg has 12 protons and therefore 12 electrons. Similarly Fe has 26 protons and 26 electrons. In an ion there is a net charge, this is caused by changing the number of electrons. If an ion has more electrons than protons it is negatively charged, and called an anion. Conversely, if the ion has fewer electrons than protons it is positively charged, and called a cation. $^{24}Mg^{2+}$, and $^{56}Fe^{2+}$ have a net positive charge of 2 meaning there are two less electrons than protons. Therefore $^{24}Mg^{2+}$, has (12-2) or 10 electrons and $^{56}Fe^{2+}$ has (26-2) or 24 electrons.

1.1 Determine the number of protons, neutrons, and electrons in the following: ^{14}N, ^{2}H, ^{81}Br, and ^{79}Br.

1.2 Determine the number of protons, neutrons, and electrons in the following: ^{199}Hg, ^{238}U, ^{124}Sn, and ^{58}Fe.

1.3 Determine the number of protons, neutrons, and electrons in the following: ^{55}Mn, ^{58}Ni, ^{60}Ni, and ^{69}Ga.

1.4 Determine the number of protons, neutrons, and electrons in the following: ^{80}Kr, ^{18}O, $^{16}O^{2-}$, and $^{19}F^{-}$.

1.5 Determine the number of protons, neutrons, and electrons in the following: ^{37}Cl, ^{23}Na, $^{37}Cl^-$, and $^{23}Na^+$

Solved problem 2: Hydrogen exists in two natural isotopes in the following proportions:

Isotope	Atomic Mass (amu.)	Abundance (wt. %)
^1H	1.007825	99.985
^2H	2.0140	0.015

Determine the atomic mass of hydrogen.

The atomic mass takes into account the relative proportions of each isotope. Thus the atomic mass will be: (0.99985)(1.007825 amu.) + (0.00015)(2.01400 amu.) = 1.00797 amu.

1.6 Lithium exists in two natural isotopes in the following proportions:

Isotope	Atomic Mass (amu.)	Abundance (wt. %)
^6Li	6.01512	7.42
^7Li	7.01600	92.58

Determine the atomic mass of lithium. (ans. 6.941 amu.)

1.7 Boron exists in two natural isotopes in the following proportions:

Isotope	Atomic Mass (amu.)	Abundance (wt. %)
^{10}B	10.0129	19.78
^{11}B	11.0931	80.22

Determine the atomic mass of boron. (ans. 10.811 amu.)

1.8 Carbon exists in two natural isotopes in the following proportions:

Isotope	Atomic Mass (amu.)	Abundance (wt. %)
^{12}C	12.0000	98.89
^{13}C	13.00335	1.11

Determine the atomic mass of carbon. (ans. 12.01115 amu.)

1.9 Magnesium exists in three natural isotopes in the following proportions:

Isotope	Atomic Mass (amu.)	Abundance (wt. %)
^{24}Mg	23.98504	78.70
^{25}Mg	24.98584	10.13
^{26}Mg	25.98259	11.17

Determine the atomic mass of magnesium. (ans. 24.312 amu.)

1.10 Chromium exists in four natural isotopes in the following proportions:

Isotope	Atomic Mass (amu.)	Abundance (wt. %)
^{50}Cr	49.9461	4.31
^{52}Cr	51.9405	83.76
^{53}Cr	52.9407	9.55
^{54}Cr	53.9389	2.38

Determine the atomic mass of Cr. (ans. 51.996 amu.)

1.11 Germanium exists in four natural isotopes in the following proportions:

Isotope	Atomic Mass (amu.)	Abundance (wt. %)
^{72}Ge	71.9217	27.43
^{73}Ge	72.9234	7.76
^{74}Ge	73.9219	36.54
^{76}Ge	75.9214	7.76

Determine the atomic mass of Ge. (ans. 72.59 amu.)

Solved Problem 3: ^{93}Nb has a rest mass of 92.906000 amu, determine the nuclear binding energy. (ans. 805 MeV/atom, 7.8×10^{13}J/mole)

The nuclear binding energy is the energy that would be given off if the nucleus were broke up into its individual particles. The strong nuclear force is large enough that there is a measurable change in mass; that is, the nucleus actually has less mass than its constituent nucleons. This energy can be expressed as,

$$E = \Delta mc^2$$
$$\Delta m = m_{nucleons} - m_{nucleus}$$

where, $m_{nucleons}$ is the net rest mass of the nucleons (protons and neutrons) and $m_{nucleus}$ is the rest mass of the nucleus. Therefore the first step is to determine the mass of the nucleus. This requires knowing the number of protons, neutrons, and electrons in the ^{93}Nb atom. Nb has atomic number 41 from the periodic table. Thus it has 41 protons, (93-41) or 52 neutrons, and 41 electrons.

The mass of the atom is the sum of the mass of the electrons and the mass of the nucleus. The mass of the nucleus is therefore,

$$
\begin{aligned}
m_{nucleus} &= m_{atom} - m_{e^-} \\
&= 92.906000 \text{ amu.} - 41(5.485803 \times 10^{-4} \text{ amu.}) \\
&= 92.88350 \text{ amu.}
\end{aligned}
$$

Knowing the mass of the nucleus we can calculate the binding energy,

$$E = \Delta mc^2$$
$$
\begin{aligned}
\Delta m &= m_{nucleons} - m_{nucleus} \\
&= (41m_p + 52m_n) - m_{nucleus} \\
&= (41 \times 1.00727647 \text{ amu.} + 52 \times 1.0086659 \text{ amu.}) - 92.88350 \text{ amu.}
\end{aligned}
$$

$$
\begin{aligned}
E &= 0.8654 \text{ amu.} \times \frac{1.66 \times 10^{-27} \text{ kg}}{1 \text{ amu.}} \times \left(3 \times 10^8 \frac{m}{s}\right)^2 \\
&= 1.2929 \times 10^{-10} \frac{J}{atom} \\
&= 1.2929 \times 10^{-10} \frac{J}{atom} \times \frac{1eV}{1.6 \times 10^{-19} J} = 8.08 \times 10^8 \frac{eV}{atom} = 808 \frac{MeV}{atom} \\
&= 1.2929 \times 10^{-10} \frac{J}{atom} \times 6.023 \times 10^{23} \frac{atoms}{mole} = 7.8 \times 10^{13} \frac{J}{mole}
\end{aligned}
$$

Note this problem could be solved more simply in two ways. First a mass defect of 1 amu correspond to a binding energy of 931.5 MeV per atom. You can calculate this from $E = \Delta mc^2$, using the information in the solution. Second Δm also equals the atomic mass of the isotope minus the net mass of electrons, protons, and neutrons. The calculation of $m_{nucleus}$ is mathematically unnecessary. However the electrons are unaffected by the strong force which holds the nucleus together. The mass defect is caused by the strong force not the electrostatic force which holds the nucleus together. To illustrate this point we broke up the calculation.

1.12 ^{35}Cl has a rest mass of 34.968850 amu, determine the nuclear binding energy. (ans. 298 MeV/atom, 2.9×10^{13} J/mole)

1.13 ^{4}He has a rest mass of 4.002600 amu, determine the nuclear binding energy. (ans. 28 MeV/atom, 2.7×10^{12} J/mole)

1.14 ^{2}H has a rest mass of 2.01400 amu, determine the nuclear binding energy. (ans. 2.3 MeV/atom, 8.2×10^{11} J/mole)

1.15 ^{69}Ga has a rest mass of 68.92570 amu, determine the nuclear binding energy. (ans. 601 MeV/atom, 5.8×10^{13} J/mole)

1.16 ^{56}Fe has a rest mass of 55.934900 amu, determine the nuclear binding energy. (ans. 506 805 MeV/atom, 4.9×10^{13} J/mole)

1.17 ^{1}H has a rest mass of 1.007970 amu, determine the nuclear binding energy. (ans. 0 MeV/atom, 0 J/mole)

PROBLEM SET #2 ➤ *Nuclear Reactions (Decay, Fission, and Fusion)*

Solved Problem 4: ^{242}Pu decays by emitting an α-particle, determine the end product, and the kinetic energy of the α-particle.

In radioactive decay the net atomic number and the net atomic mass number is conserved. In the case of α-decay the net number of protons and neutrons are conserved. For radiation problems it is convenient to add the atomic number as a prefix to the element symbol. Thus the α-particle is $^4_2$He, and ^{242}Pu is $^{242}_{94}$Pu. The reaction can be written as,

$$^{242}_{94}\text{Pu} \rightarrow\ ^4_2\text{He}\ +\ ^a_b\text{X}$$

Since the net atomic mass number is conserved, $242 = 4 + a$, $a = 238$. Similarly the net atomic number is conserved so $94 = 2 + b$, $b = 92$. Since the atomic number of the product element X is 92, it is U from the periodic table. The final products are the α particle and ^{238}U.

The mass of the products will be less than the mass of the reactants; this release of energy will be given up to the α particle as kinetic energy. This energy is,

$$E = \Delta mc^2$$
$$\Delta m = m_{Pu} - (m_\alpha + m_U)$$
$$= 242.0587 \text{ amu.} - (4.00151 \text{ amu.} + 238.050800 \text{ amu.})$$
$$= 0.00639 \text{ amu.}$$
$$E = 0.00639 \text{ amu} \times 931.5 \frac{\text{MeV}}{\text{amu}}$$
$$= 5.95 \text{ MeV}$$

Note that the conversion factor $931.5 \frac{\text{MeV}}{\text{amu.}}$ is found by calculating mc^2 for a mass of 1 amu.

2.1 ^{234}U decays by emitting an α-particle, determine the end product, and the kinetic energy of the α-particle. (ans. ^{230}Th, 5.85 MeV)

2.2 ^{222}Rn decays by emitting an α-particle, determine the end product, and the kinetic energy of the α-particle. (ans. ^{218}Po, 6.60 MeV)

2.3 ^{210}Po decays by emitting an α-particle, determine the end product, and the kinetic energy of the α-particle. (ans. ^{206}Pb, 6.42 MeV)

2.4 ^{238}U decays by emitting an α-particle, determine the end product, and the kinetic energy of the α-particle. (ans. ^{234}Th, 5.30 MeV)

2.5 ^{3}H decays by emitting a β-particle, determine the end product, and the maximum kinetic energy of the β-particle. (ans. ^{3}He, 0.018 MeV)

2.6 ^{234}Pa decays by emitting a β-particle, determine the end product, and the maximum kinetic energy of the β-particle. (ans. ^{234}U, 0.027 MeV)

Solved Problem 6: ^{218}Po decays by emitting a β-particle; determine the end product, and the maximum kinetic energy of the β-particle. (ans. ^{218}At, 0.255 MeV)

A β-particle is an electron. During a β decay a neutron becomes a proton and an electron is emitted from the nucleus into the electron cloud. To account for this the β particle is usually written as follows, $^{0}_{-1}\beta$. This accounts for the change in atomic number and no change in atomic mass. However we can identify the end product by assuming the net atomic number and net atomic mass number remains constant. Thus the reaction is,

$$^{218}_{84}\text{Po} \rightarrow\ ^{0}_{-1}\beta\ +\ ^{a}_{b}\text{X}$$

a = 218, and b-1=84 or b=85. The atomic element with atomic number 85 is At. The final products therefore are $^{218}_{85}$At and $^{0}_{-1}\beta$.

The mass of the products will be less than the mass of the reactants; this release of energy will be given up to the a particle as kinetic energy. This energy is,

$$E = \Delta mc^2$$
$$\Delta m = m_{Po} - (m_\beta + m_{At})$$
$$= 218.008969\ \text{amu.} - (0\ \text{amu.} + 218.008695\ \text{amu.})$$
$$= 2.73 \times 10^{-4}\ \text{amu.}$$
$$E = 2.73 \times 10^{-4}\ \text{amu.} \times 931.5\ \frac{\text{MeV}}{\text{amu.}}$$
$$= 5.95\ \text{MeV}$$

Note that the conversion factor, $931.5\ \frac{\text{MeV}}{\text{amu.}}$, is found by calculating mc^2 for a mass of 1 amu.

Solved Problem 7: ^{235}U when bombarded by a single neutron undergoes fission to produce ^{93}Rb and ^{141}Cs. a) Write the balanced reaction. b) Is this a chain reaction? c) How much energy is given off from this reaction? d) How much energy is given off per mole of reactant? e) How much energy is given off per 100g of reactant?

a) As in all nuclear reactions the net atomic number and the net atomic mass number must be conserved. The atomic number for a neutron is 0 and a neutron is usually written as, $^{1}_{0}$n. In fission reactions the products are other elements and neutrons. The atomic number for Cs is 55, for Rb is 37 and for U is 92. Thus the net atomic number is conserved. The only product will be the two elements and neutrons. The net reaction is,

$$^{235}U + {}^{1}_{0}n \rightarrow {}^{93}Rb + {}^{141}Cs + X\ {}^{1}_{0}n,$$

balancing the net atomic mass number gives, $236 = 93 + 141 + X$, $X = 236 - (93 + 141) = 2$. The final reaction is therefore,

$$^{235}U + {}^{1}_{0}n \rightarrow {}^{93}Rb + {}^{141}Cs + 2\ {}^{1}_{0}n,$$

b) The reaction gives off 2 neutrons; these neutrons can bombard other ^{235}U nuclei, therefore it is a chain reaction.

c) The energy given off is calculated as the amount of mass converted to energy. That is the products will actually be lighter than the reactants.

$E = \Delta mc^2$

$\Delta m = m_{U\text{-}235+m^n} - (m_{Rb\text{-}98} + 3m_n)$

$\quad = (235.043900\ \text{amu.} + 1.0086659\ \text{amu.}) - (92.92171766\ \text{amu.} +$

$\qquad 140.9194847\ \text{amu.} + 2(1.0086659\ \text{amu.}))$

$\quad = 2.73 \times 10^{-4}\ \text{amu.}$

$E = 0.194\ \text{amu.} \times 931.5\dfrac{\text{MeV}}{\text{amu.}}$

$\quad = 5.95\ \text{MeV}$

d) Thus each fission produces 180.7 MeV of energy. To calculate the energy produced per mole of ^{235}U, we simply convert,

$$E = 180.7 \times 10^6 \frac{eV}{atom} \times 1.6 \times 10^{-19} \frac{J}{eV} \times 6.023 \times 10^{23} \frac{atoms}{mole}$$

$$= 1.78 \times 10^{13} \frac{J}{mole}$$

e) To estimate the amount of energy produced by 100g of ^{235}U we use the molecular weight of approximately 235 g/mole.

$$E = 1.78 \times 10^{13} \frac{J}{mole} \times \frac{235 \text{ g/mole}}{100 \text{ g.}}$$

$$= 7.6 \times 10^{12} \frac{J}{100 \text{ g.}}$$

So 100 g of ^{235}U when fissioned in this manner will produce 7.6×10^{12} J of energy.

2.7 ^{235}U when bombarded by a single neutron undergoes fission to produce ^{105}Mo and ^{128}Sn. a) Write the balanced reaction. b) Is this a chain reaction? c) How much energy is given off from this reaction? d) How much energy is given off per mole of reactant? e) How much energy is given off per 100g of reactant?

(ans. ^{235}U $+ {}^{1}_{0}$n $\rightarrow {}^{105}$Mo $+ {}^{128}$Sn $+ 3 {}^{1}_{0}$n, Yes it is a chain reaction, 185.3 MeV/atom, 1.78×10^{13} J/mol, 7.6×10^{12} J/100g.)

2.8 ^{233}U when bombarded by a single neutron undergoes fission to produce ^{92}Sr and ^{138}Xe. a) Write the balanced reaction. b) Is this a chain reaction? c) How much energy is given off from this reaction? d) How much energy is given off per mole of reactant? e) How much energy is given off per 100g of reactant?

(ans. ^{233}U $+ {}^{1}_{0}$n $\rightarrow {}^{92}$Sr $+ {}^{138}$Xe $+ 4 {}^{1}_{0}$n, Yes it is a chain reaction, 175.7 MeV/atom, 1.69×10^{13} J/mole, 7.3×10^{12} J/100g.)

2.9 ^{239}Pu when bombarded by a single neutron undergoes fission to produce ^{100}Mo and ^{140}Xe. a) Write the balanced reaction. b) Is this a chain reaction? c) How much energy is given off from this reaction? d) How much energy is given off per mole of reactant? e) How much energy is given off per 100g of reactant?

(ans. ^{239}Pu $+ {}^{1}_{0}$n $\rightarrow {}^{100}$Mo $+ {}^{140}$Xe , No it is not a chain reaction, 212 MeV/atom, 3.39×10^{-11} J/mol, 1.4×10^{-11} J/100g.)

Solved Problem 7: ^{4}He is formed from ^{2}H through nuclear fusion. a) Write the balanced reaction. b) How much energy is given off by the reaction? c) How much energy is given off per mole of product formed? d) How much energy is given off per 100 g of reactant consumed?

a) Using the periodic table we can write the atomic number in as a subscript preceding the isotope. Thus ^{4}He is written as $^4_2$He, and ^{2}H is written as $^2_1$H. Balancing the net atomic numbers and net atomic mass numbers, the fusion reaction is,

$$2\,{}^2_1\text{H} \rightarrow {}^4_2\text{He},$$

b) The energy of the reaction is calculated as a mass defect. That is the net mass of the reactants are heavier than the net mass of the products.

$$E = \Delta mc^2$$

$$\Delta m = 2m_H - m_{He}$$

$$= 2(2.01400 \text{ amu.}) - 4.002600$$

$$= 0.0254 \text{ amu.}$$

$$E = (0.254 \text{ amu.})\,\frac{931.5 \text{ MeV}}{\text{amu.}}$$

$$= 23.6 \text{ MeV}$$

Thus for each He atom produced 23.6 MeV of energy are released.

c) The energy per mole of product formed is,

$$E = 23.6 \times 10^6\,\frac{eV}{\text{atom}} \times 1.6 \times 10^{-19}\,\frac{J}{eV} \times 6.023 \times 10^{23}\,\frac{\text{atoms}}{\text{mole}}$$

$$= 1.14 \times 10^{12}\,\frac{J}{\text{mole}}$$

d) In this case there are two moles of reactant for every mole of product consumed. Thus for each mole of ^{2}H consumed 5.7×10^{11} J are given off. Thus the amount of energy given off per 100g of ^{2}H is,

$$E = 5.7 \times 10^{11}\,\frac{J}{\text{mole}} \times 100\text{g}.\frac{2.014 \text{ g}/\text{mole}}{2.014 \text{ g}/\text{mole}}$$

$$= 2.83 \times 10^{13}\,\frac{J}{100\text{g}.}$$

2.10 ^{28}Si is formed from ^{12}C and ^{16}O through nuclear fusion. a) Write the balanced reaction. b) How much energy is given off by the reaction? c) How much energy is given off per mole of product formed? d) How much energy is given off per 100 g of reactant consumed?

(ans. ^{12}C + ^{16}O → ^{28}Si , 16.75 MeV, $1.61×10^{12}$ J/mol, $1.0×10^{12}$ J/100g)

2.11 ^{32}S is formed from ^{16}O through nuclear fusion. a) Write the balanced reaction. b) How much energy is given off by the reaction? c) How much energy is given off per mole of product formed? d) How much energy is given off per 100 g of reactant consumed?

(ans. 2 ^{16}O → ^{32}S , 16.5 MeV, $1.59×10^{12}$ J/mol, $4.97×10^{12}$ J/100g)

2.12 ^{12}C is formed from ^{4}He through nuclear fusion. a) Write the balanced reaction. b) How much energy is given off by the reaction? c) How much energy is given off per mole of product formed? d) How much energy is given off per 100 g of reactant consumed?

(ans. 3 ^{4}He → ^{12}C , 7.2 MeV, $6.99×10^{11}$ J/mol, $5.83×10^{12}$ J/100g)

2.13 ^{3}He is formed from ^{2}H and ^{1}H through nuclear fusion. a) Write the balanced reaction. b) How much energy is given off by the reaction? c) How much energy is given off per mole of product formed? d) How much energy is given off per 100 g of reactant consumed?

(ans. ^{2}H + ^{1}H → ^{3}He , 5.5 MeV, $5.3×10^{11}$ J/mol, $2.66×10^{13}$ J/100g)

PROBLEM SET #3 ➤ *Basic Quantum Mechanics*

Solved Problem 8:   Determine the energy of a light wave with a wavelength of 430 nm.

Planck's observation for electromagnetic radiation was E=hv, where v is the frequency and h is Planck's constant ($6.63×10^{-34}$ J·sec., or $4.144×10^{-15}$ eV·sec.). Similarly for any wave the product of the frequency and the wavelength is the velocity of the wave.

Therefore Planck's observation can be rewritten as, $E = \dfrac{hc}{\lambda}$ where C is the velocity of light (3.0×10^8 m/s) and λ is the wavelength.

The energy of a light wave with a wavelength of 430 nm. is,

$$E = \frac{hc}{\lambda}$$
$$= \frac{\left(4.144 \times 10^{-15}\,\text{EV·sec.}\right)\left(3 \times 10^8\,\text{m/sec.}\right)}{430 \times 10^{-9}\,\text{m}}$$
$$= 2.89\,\text{eV}$$

3.1 Determine the energy of a light wave with a wavelength of 690 nm. (ans. 1.80 eV)

3.2 Determine the energy of a light wave with a wavelength of 550 nm. (ans. 2.26 eV)

3.3 Determine the energy of an X-ray wave with a wavelength of 0.154 nm. (ans. 8.1 keV)

3.4 Determine the energy of a light wave with a wavelength of 0.077 nm. (ans. 16.1 keV)

3.5 Determine the frequency of a light wave with a wavelength of 600 nm. (ans. 2.08 eV)

3.6 Determine the frequency of an ultraviolet wave with a wavelength of 200 nm. (ans. 6.21 eV)

3.7 Determine the frequency of an infrared light wave with a wavelength of 800 nm. (ans. 1.55 eV)

3.8 Determine the frequency of a radio wave with a wavelength of 1 m. (ans. 1.2×10^6 eV)

3.9 The first ionization energy of Li is 5.39 eV, what wavelength is required to ionize Li? (ans. 230 nm.)

3.10 The first ionization energy of Al is 5.9 eV, what wavelength is required to form Al^+? (ans. 210 nm.)

3.11 The first ionization energy of Sn is 7.3 eV, what wavelength is required to ionize Sn^+? (ans. 170 nm.)

3.12 The first ionization energy of Pt is 9.0 eV, what wavelength is required to ionize Pt ? (ans. 138 nm.)

3.13 The first ionization energy of Na is 5.13 eV, what is wavelength is required to ionize Li? (ans. 242 nm.)

Solved Problem 9: Consider the plot for a photoelectric effect experiment on K, Determine the work function of K. What is the energy of an ejected electron under incident light of 4×10^{15} Hz, 2×10^{16} Hz?, and 1×10^{14} Hz?

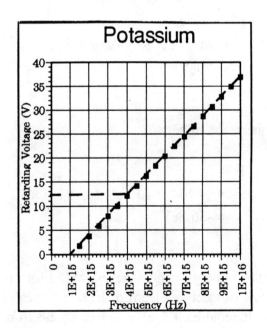

The work function will be the energy required to just eject an electron. In the photoelectric effect the retarding voltage represents the voltage needed to stop electron motion. The electron volt is defined as the energy of an electron as it passes through a voltage of 1V. Therefore if the retarding voltage is 20V, then 20 eV of energy are required to stop electron motion. Thus the electrons would have a kinetic energy of 20 eV. To find the work function we can extrapolate the data to the x-axis.

Doing this we see that the x-axis intercept is at 1.0×10^{15} Hz. The work function therefore corresponds to the energy of this wave.

$$E = h\nu$$
$$= 4.144 \times 10^{-15} \text{eV} \cdot \text{sec.} \times 1 \times 10^{15} \text{sec}^{-1}$$
$$= 4.144 \text{eV}$$

To determine the energy of electrons under incident light of 4×10^{15} Hz, we can read the retarding voltage of the graph. At a frequency of 4×10^{15} Hz it requires 12.5 V to stop electron flow. Thus the energy of the electrons is 12.5 eV.

The frequency 2×10^{16} Hz is not on the graph. However we know the work function is 4.1 eV, and the slope of the line is h, so we can calculate the energy as,

$$E = h\nu - \phi$$
$$= (4.144 \times 10^{-15} \text{eV} \cdot \text{sec.})(2.0 \times 10^{16} \text{sec}^{-1}) - 4.1 \text{eV}$$
$$= 79 \text{eV}$$

The frequency, is not on the graph either so,

$$E = h\nu - \phi$$
$$= (4.144 \times 10^{-15} \text{eV} \cdot \text{sec.})(1.0 \times 10^{14} \text{sec}^{-1}) - 4.1 \text{eV}$$
$$= -36.8 \text{eV}$$

This is impossible; the kinetic energy cannot be ejected. From the graph we can see that this frequency is too low to eject an electron. In fact waves of this frequency only have an energy of 0.4 eV. It requires 4.1 eV to eject an electron, so no electrons are ejected. The kinetic energy is therefore 0.

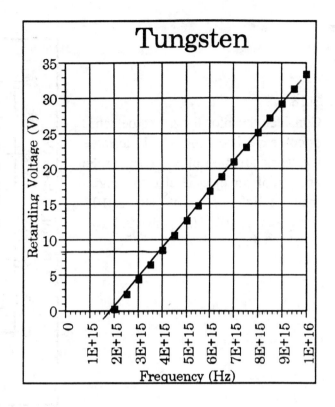

3.14

Consider the plot for a photoelectric effect experiment on W, Determine the work function of W. What is the energy of an ejected electron under incident light of 4×10^{15} Hz, 2×10^{16} Hz?, and 1×10^{14} Hz? (ans. 7.9 eV, 9 eV, 80 eV, 0 eV)

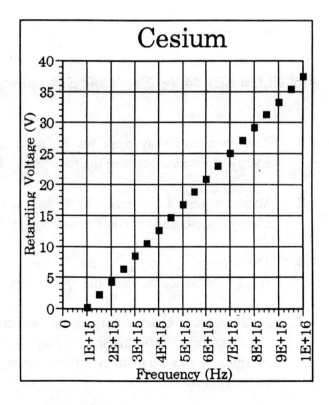

3.15

Consider the plot for a photoelectric effect experiment on Fe, Determine the work function of Cs. What is the energy of an ejected electron under incident light of 4×10^{15} Hz, 2×10^{16} Hz?, and 1×10^{14} Hz? (ans. 3.8 eV, 85 eV, 0 eV)

3.16

Consider the plot for a photoelectric effect experiment on Pt, Determine the work function of Pt. What is the energy of an ejected electron under incident light of 4×10^{15} Hz, 2×10^{16} Hz?, and 1×10^{14} Hz? (ans. 9.0 eV, 78 eV, 0 eV)

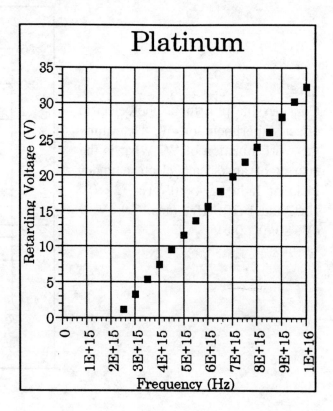

PROBLEM SET #4 ➤ *Electronic Structure and Configuration*

4.1 The strongest X-ray peak for B has a wavelength of 6.75 nm? What is the energy difference between the two energy levels in B?

4.2 The strongest X-ray peak for C has a wavelength of 4.47 nm? What is the energy difference between the two energy levels in C?

4.3 The strongest X-ray peak for O has a wavelength of 2.36 nm? What is the energy difference between the two energy levels in O?

4.4 Ne shows two characteristic X-ray wavelengths 1.44 nm and 1.46 nm. These represent electrons in the second shell dropping into the first shell. What are the energies of these transitions? How closely spaced are the two energy levels in the second shell?

Solved Problem: Sodium has four characteristic X-ray wavelengths? (1.15 nm, 1.19 nm, 37.5 nm, and 40.6 nm.) Characterize these transitions.

Each of these transitions corresponds to an energy difference. The highest energy differences correspond to electrons falling from higher shells to the **lowest energy shell**. Note there are two of these, the two lowest wavelengths. Using the formula, $E = \frac{hc}{\lambda}$, we can see that the energies of these transitions are 1.071 keV and 1.041 keV respectively. These are electrons from the second shell dropping into the first shell.

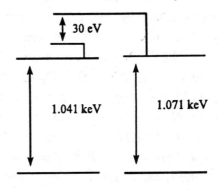

This shows that there are 2 very closely spaced energy levels in the second shell which differ by about 30 eV. The other two energies are 30 eV and 33 eV respectively indicating a transition between these two levels. Thus from this information we can characterize the energy levels in Na as shown.

Note these transitions are between the 2p and the 1s, 2s and 1s, and 2p and 2s shells. The outermost 3s electron may play a role in one of the higher transitions. This does show that there are two distinct energy levels, and that within the upper level there is a distinct split. The 2nd level has two subshells.

You should be cautioned that these data are not sufficient to completely characterize the element. Not all electrons participate in this and this does not give a complete description of the electron energy structure. It does however show that there are distinct energy levels, and distinct sublevels.

Solved Problem: Potassium shows seven characteristic X-ray wavelengths (0.343 nm., 0.345 nm, 0.373 nm, 0.374 nm, 4.72 nm, 4.77 nm, and 69.2 nm.) Characterize these transitions.

Each of these transitions is caused by an electron dropping from a high energy level into a lower energy level. Using the formula, $E = \frac{hc}{\lambda}$, we can see that the energies of these transitions are: 3.60 keV, 3.58 keV, 3.31 keV, 3.31 keV, 263 eV, 259 eV, and 17.9 eV. These seven energies indicate that there is a transition of about 3600eV between one shell and the lowest shell, a transition of about 3300 eV between another shell and the lowest shell, and a transition of about 260 eV between two shells.

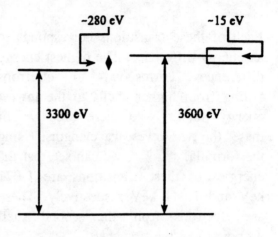

The 260 eV difference seems to correspond to the difference between the two higher shells and the lowest shell. This indicates the presence of three distinct shells of electrons. The 17 eV transition corresponds to a split in the outermost shell of about 17 eV. The split in the second shell is less pronounced. We can therefore characterize the electron as having a lowest shell, a shell approximately 3300 eV higher, and a third shell approximately 300 eV higher than the second shell. The second shell shows two levels of slightly different energy and the third level shows two levels approximately 17 eV apart.

You should be cautioned that these data are not sufficient to completely characterize the element. Not all electrons participate in this and this does not give a complete description of the electron energy structure. It does however show that there are distinct energy levels, and distinct sublevels.

4.5 Ca has shows nine distinct X-ray wavelengths: 0.307 nm, 0.3087 nm, 0.3356 nm, 0.3359 nm, 3.5925 nm, 3.6304 nm, 4.044 nm, 4.093 nm, and 52.5 nm. Characterize the shell structure of Ca from these data.

4.6 Co shows 10 distinct X-ray wavelengths: 0.1607 nm, 0.1620 nm, 0.1788 nm, 0.1792 nm, 1.424 nm, 1.565 nm, 1.785 nm, 1.828 nm, and 21.363 nm. Characterize the shell structure of Co from these data.

4.7 Cl shows five distinct X-ray wavelengths: 0.440 nm, 0.4724 nm., 0.4727 nm., 6.73 nm, and 6.79 nm. Characterize the shell structure of Cl from these data.

4.8 Cr shows 10 distinct X-ray wavelengths: 0.206 nm, 0.208 nm., 0.227 nm., 0.228 nm., 1.89 nm, 21.2 nm., 21.6 nm., 24.3 nm., 24.7 nm., and 30.9 nm. Characterize the shell structure of Cr from these data.

4.9 Fe shows 10 distinct X-ray wavelengths: 0.174 nm., 0.176 nm., 0.1935 nm., 0.1939 nm., 1.564 nm., 1.725 nm., 1.758 nm., 1.973 nm., 2.014 nm., 24.29 nm. Characterize the shell structure of Fe from these data.

4.10 Si shows four distinct X-ray wavelengths: 0.675 nm., 0.7121 nm., 0.7124 nm., and 13.54 nm. Characterize the shell structure of Si from these data.

Solved Problem: Write the electron configuration for: He, Be, C, Na, Fe, Fe^{2+}, and Br^-.

The first step is to determine the number of electrons in each of these elements. Form the periodic table He has an atomic number of 2, Be 4, C 6, Na 11, Fe 26, and Br 35. The atomic number indicates the number of protons in the nucleus. Next we need to calculate the number of electrons in the species. For a neutral atom the number of electrons equals the number of protons. Thus He has 2 electrons, Be 4, C 6, Na 11, and Fe 26. Ions have a net charge because the number of electrons is either greater than or less than the number of protons in the nucleus. Fe^{2+} has a net positive charge of 2, and therefore there are two fewer electrons than protons. Fe^{2+} has 24 electrons. Similarly Br^- has a net negative charge of 1, and has one additional electron, giving it 36 electrons.

Once we have determined the number of electrons; we can determine the electron configuration by filling the lowest energy orbital first. Note the filling order is 1s, 2s, 2p, 3s, 3p, 4s, 3d, 4p, 5s, 4d, 5p, 6s, 4f, 5d, 6p, 7s, 5f, and 6d... . The pattern would continue but this includes all known elements to date. Note also that there is only one s orbital, there are three p orbitals, five d orbitals and seven f orbitals. This means that the s sublevel can hold two electrons, the p 6 electrons, the d 10 electrons and the f 14 electrons.

He -2 electrons: The lowest energy level is 1s which has a capacity of 2 electrons. Placing two electrons in this level we have placed all electrons. The electron configuration for He is: $1s^2$.

Be - 4 electrons: The lowest energy level is 1s which has a capacity of 2 electrons. Placing two electrons in this level we have 2 more electrons to place. The next level is the 2s which has a capacity of 2 electrons. Placing two electrons in this level, we have placed all electrons. The electron configuration for Be is: $1s^2 2s^2$.

C - 6 electrons: The lowest energy level is 1s which has a capacity of 2 electrons. Placing two electrons in this level we have 4 more electrons to place. The next level is the 2s which has a capacity of 2 electrons, leaving 2 more electrons to place. The next energy level is the 2p which has a capacity of 6 electrons. Placing two electrons in this level, we have placed all electrons. The electron configuration for C is: $1s^2 2s^2 2p^2$.

Na - 11 electrons: The lowest energy level is 1s which has a capacity of 2 electrons. Placing two electrons in this level we have 9 more electrons to place. The next level is the 2s which has a capacity of 2 electrons, leaving 7 more electrons to place. The next energy level is the 2p which has a capacity of 6 electrons, leaving 1 more electron to place. The next energy level is the 3s which has a capacity of 2 electrons Placing 1 electron in this level, we have placed all electrons. The electron configuration for Na is: $1s^2 2s^2 2p^6 3s^1$.

Fe- 26 electrons: The lowest energy level is 1s which has a capacity of 2 electrons. Placing two electrons in this level we have 24 more electrons to place. The next level is the 2s which has a capacity of 2 electrons, leaving 22 more electrons to place. The next energy level is the 2p which has a capacity of 6 electrons, leaving 16 more electron to place. The next energy level is the 3s which has a capacity of 2 electrons, leaving 14 more electrons to place. The next energy level is the 3p with a capacity of 6 electrons, leaving 8 electrons to place. The next energy level is the 4s with a capacity of 2 electrons, leaving 6 electrons to place. The next level is the 3d with a capacity of 10 electrons Placing 6 electrons in this level, we have placed all electrons. The electron configuration for Fe is: $1s^2 2s^2 2p^6 3s^2 3p^6 4s^2 3d^6$.

Fe^{2+}- 24 electrons: The lowest energy level is 1s which has a capacity of 2 electrons. Placing two electrons in this level we have 22 more electrons to place. The next level is the 2s which has a capacity of 2 electrons, leaving 20 more electrons to place. The next energy level is the 2p which has a capacity of 6 electrons, leaving 14 more electrons to place. The next energy level is the 3s which has a capacity of 2 electrons, leaving 12 more electrons to place. The next energy level is the 3p with a capacity of 6 electrons, leaving 6 electrons to place. The next energy level is the 4s with a capacity of 2 electrons, leaving 4 electrons to place. The next level is the 3d with a capacity of 10 electrons Placing 4 electrons in this level, we have placed all electrons. The electron configuration for Fe^{2+} is: $1s^2 2s^2 2p^6 3s^2 3p^6 4s^2 3d^4$.

Br^- - 36 electrons: The lowest energy level is 1s which has a capacity of 2 electrons. Placing two electrons in this level we have 34 more electrons to place. The next level is the 2s which has a capacity of 2 electrons, leaving 32 more electrons to place. The next energy level is the 2p which has a capacity of 6 electrons, leaving 26 more electrons to place. The next energy level is the 3s which has a capacity of 2 electrons, leaving 24 more electrons to place. The next energy level is the 3p with a capacity of 6 electrons, leaving 18 electrons to place. The next energy level is the 4s with a capacity of 2 electrons, leaving 16 electrons to place. The next level is the 3d with a capacity of 10 electrons, leaving 6 electrons to place. The next level is the 4p with a capacity of 6 electrons. Placing 6 electrons in this level, we have placed all electrons. The electron configuration for Br^- is: $1s^2 2s^2 2p^6 3s^2 3p^6 4s^2 3d^{10} 4p^6$.

4.11   Write the electron configuration for: Sn, F, Be, and Na.

4.12   Write the electron configuration for: Mg, Mn, B, and C.

4.13   Write the electron configuration for: Rb, Cs, Po, and Yb.

4.14   Write the electron configuration for: W, Cr, Mo, and Fe.

4.15   Write the electron configuration for: Au, Cu, Ag, and Pt.

4.16   Write the electron configuration for: H^+, H, and H^-.

4.17   Write the electron configuration for: O^{+2}, O, and O^{-2}.

4.18   Write the electron configuration for: Cl^-, F^-, Cl and F.

4.19   Write the electron configuration for: Na^+, K^+, Na and K.

4.20   Write the electron configuration for: Ca^{+2}, Fe^{+2}, Fe^{+3}, and Al^{+3}.

Solved Problem: In each of the following determine the number of valence electrons, and the number of unpaired electrons: He, Be, C, Na, Fe, Fe^{2+}, and Br^-.

The first step is to determine the electron configuration. This was done in the previous solved problem. The electron configurations are: He - $1s^2$, Be - $1s^22s^2$, C - $1s^22s^22p^2$, Na - $1s^22s^22p^63s^1$, Fe - $1s^22s^22p^63s^23p^64s^23d^6$, Fe^{2+} - $1s^22s^22p^63s^23p^64s^23d^4$, and Br^- - $1s^22s^22p^63s^23p^64s^23d^{10}4p^6$.

The valence electrons and the unpaired electrons will be in the outermost shell.

He- $1s^2$ - The outermost shell corresponds to n=1. There are therefore 2 valence electrons. There is only one 1s orbital since it contains two electrons, there are no unpaired electrons.

Be- $1s^22s^2$ - The outermost shell corresponds to n=2. There are therefore 2 valence electrons. There is only one 2s orbital since it contains two electrons, there are no unpaired electrons.

C - $1s^22s^22p^2$ - The outermost shell corresponds to n=2, there are two electrons in the 2s orbitals and two electrons in the 2p orbitals, giving four electrons in the valence shell. There is only 1 2s orbital so these two electrons are paired, however there are 3 2p orbitals so these two are unpaired.

4.21 In each of the following determine the number of valence electrons, and the number of unpaired electrons: H, He, Na, and Mg.

4.22 In each of the following determine the number of valence electrons, and the number of unpaired electrons: Fe, Cu, Mg, and Mn.

4.23 In each of the following determine the number of valence electrons, and the number of unpaired electrons: Mg, F, Ar, and O.

4.24 In each of the following determine the number of valence electrons, and the number of unpaired electrons: B, N, C, and S.

4.25 In each of the following determine the number of valence electrons, and the number of unpaired electrons: F^-, Al^{+3}, O^{2-}, and He.

PROBLEM SET #5 ➤ *Thought Questions*

5.1 What are the roles of protons, neutrons, and electrons in ion formation?

5.2 Which appears to be a more attractive energy source, nuclear fission type reactions, or nuclear fusion type reactions? Which gives off more energy, which has a higher energy barrier?

5.3 In problems 2.16 through 2.20 we discussed several reactions that take place within stars. As a star burns it eventually becomes an iron core through the fusion of other elements. What happens to the temperature of the star as it burns? Is it supported by the results of these calculations.

5.4 X-rays are generated by knocking a low level electron out of a metal and allowing higher level electrons to cascade downwards. In Cu there are two energies very close to one another. Why?

3

THE PERIODIC TABLE

"Of all the tools available to materials scientists, the most indispensable is probably a piece of paper known as the periodic table of the elements."

from "The Path of No Resistance: The Story of the Revolution in Superconductivity," Bruce Schechter, Touchstone Books, New York, 1989, p. 49.

3.1 ➤ OVERVIEW OF THE PERIODIC TABLE

We turn again to Schechter's book (pp. 49-50) for a concise summary:

"Late in the 19th century the Russian chemist Dimitri Mendeleev noticed that when the known elements were arranged in order of increasing atomic number, that is, the number of electrons that swarm around their nuclei, their chemical and physical properties recurred periodically. He made a chart of the elements that was arranged so that the periods fell neatly into columns. In order to do this, Mendeleev found that he had to leave certain gaps, holes in which he believed would fit elements yet to be discovered. Since then all Mendeleev's holes have been filled, and the observed periodicity has been explained by the quantum theory. The electrons in an atom are organized in a series of concentric shells, each of which can contain only a certain, small number of electrons. The chemical properties of an atoms are determined by the number of electrons in its outer shell; atoms with the same number of outer electrons, although they will differ in weight and size, will have similar chemical properties. Elements with the same number of outer electrons occur in the same column of the periodic table. The periodic table is a concise and orderly guide to the properties of the chemical elements."

This nice synopsis contains what appears to be contradictions. It was said that quantum theory can account for the arrangement of the periodic table, and this is true. However, it was also said that 'the electrons in an atom are organized in a series of concentric shells.' The shell model with its electron orbits is technically not correct—we should really think of orbitals, which from quantum mechanics can be defined as a 3-D shape in which the probability of finding an electron is high (say, 95%). This, however, is difficult to visualize. The shell model is appealing because most of us can imagine electron orbits much more easily than orbitals. Fortunately, the shell model is of great value - not only because we can relate more easily to it, but also because it actually has predictive value. It seems odd that a model which is technically wrong can still be useful. This is indeed the case here, and it can be a powerful tool provided that we don't exceed its limitations. Let's explore the shell model a bit further.

3.2 ➤ THE SHELL MODEL AND SOME PREDICTIONS

Consider the following sketch of a hydrogen atom using the shell model. The electron can be viewed as orbiting about the proton (nucleus). The electron should occupy the n=1 energy level and therefore this is the 1s orbital in quantum mechanical terms.

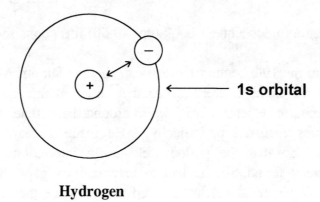

Hydrogen

The electron and nucleus can be considered as charged particles, and their attraction for each other should be governed by **Coulomb's Law**, which can be written in general as:

$$F = (Ze^+)(e^-)/r^2$$

where F is the force of attraction between the nucleus and a particular electron, Ze^+ is charge of the nucleus (e^+ is the charge of a proton and Z is the number of protons), e^- is the charge of an electron, and r is the distance between the nucleus and the electron in question. (Z=1 for the hydrogen atom.) Absorption of light of the correct wavelength can promote this electron into the next level, 2s, and here the force of attraction should be less because the distance (r) has increased. The higher energy electron will then drop to ground state, emitting a photon, because the electron is more stable in the ground state. According to Coulomb's Law, the force of attraction is greater in the ground state.

Now consider a slightly more complex atom, boron. Following the orbital filling rules introduced in the previous lecture, B can be depicted using shells as shown below.

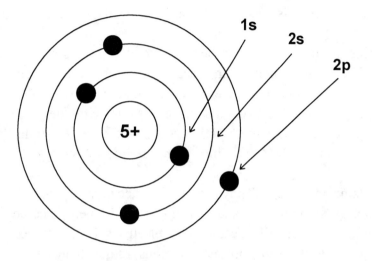

Boron

Since the size of an atom is usually determined by the radius of the outermost filled shells, a B atom is larger than H. The 1s electrons in B are attracted more strongly to the nucleus than the 1s electrons in H because they feel an effective charge of +5. Electrons further out do not, however, feel the same nuclear charge. The 2s electrons feel a charge of about +3, since the 1s electrons shield (screen) the 2s from the full nuclear charge. The 2p electrons feel a nuclear charge of only about +1 since they are screened by the 1s and 2s electrons. Therefore, the 2p electrons in B should see about the same nuclear charge as the 1s electron in H but should be easier to ionize because of the larger shell radius.

The shell model is also useful in predicting changes in atom size which occur on ionization. For example, it is found experimentally that a B^+ ion is smaller than a neutral B atom. Since size depends on the radius of the highest energy shell containing an electron, B^+ should be smaller. It has no 2p electron, and therefore its size is determined by the radius of the 2s shell, which should be smaller than that of the 2p shell.

3.3 ➤ X-RAYS

X-rays consist of electromagnetic radiation of very short wavelength (10^{-8} to 10^{-10} m) and thus high energy. You are all familiar with the medical applications of x-rays. We will see later in the course that the diffraction of x-rays provides the most compelling evidence for the existence of long range order in solids.

The generation of x-rays can be understood from Figure 1 and the shell model of the atom. X-rays are generated in practice by bombarding a metal target, held at a positive potential, with electrons. The latter are generated from a hot filament. Some of the impinging electrons make glancing collisions with the target atoms, losing a bit of energy and increasing the kinetic energy of the target atoms (and therefore the target gets hot). Some make 'direct hits' and are rapidly decelerated.

A fundamental idea of physics is that accelerating (or decelerating) charges emit electromagnetic radiation, and that in fact occurs here. The emission of electromagnetic radiation by the rapidly decelerating electrons is in essence an inverse photoelectric effect in that electrons produce photons rather than vice versa. Other impinging electrons may, if their energies are high enough, actually eject some electrons from Cu atoms. Let's imagine this happens to a 1s electron. This would create at a given instant a half-filled orbital.

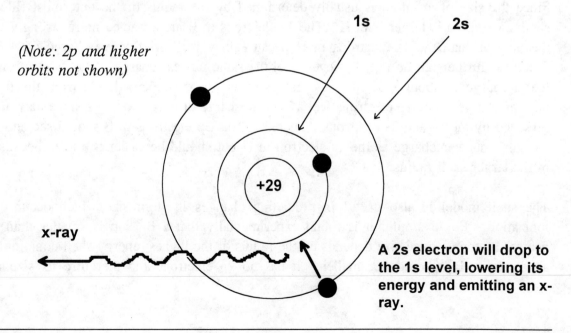

(Note: 2p and higher orbits not shown)

1s   **2s**

+29

x-ray

A 2s electron will drop to the 1s level, lowering its energy and emitting an x-ray.

The ΔE between the 1s and 2s levels for metals is such that the frequency ($\Delta E = h\nu$) happens to fall in the x-ray region of electromagnetic radiation.

As an aside, the x-rays generated from the Cu atom in the manner shown above are called K_α x-rays. The 1s shell has traditionally been called the 'K-shell,' and the $n=2$ shell (s and p) has been called the L-shell, etc. (Rumor has it that Bohr named the shells K, L, M, etc., instead of A, B, C, etc., because he wanted to leave some room in case even lower energy shells were discovered.) Since the x-ray was generated by an electron dropping to the K shell, it is referred to a K radiation. The α designates the fact that the electron dropped down from the next highest level (L-shell). If the electron dropped from the M shell ($n=3$) to K, the x-rays generated would be called K_β.

Because x-ray wavelengths are characteristic of a particular element, x-ray emission can be used to identify an unknown element. This is an important technique used almost daily at RPI and around the world. The characteristic x-ray emissions are also of significance regarding the periodic table. In 1913, the same year that Bohr announced his model of the hydrogen atom, Henry Moseley reported on results of x-ray K_α emissions from a series of metal targets. He found that a plot of $(1/\lambda_\alpha)^{1/2}$ vs. atomic number of the target element increased with atomic number and was linear. Recall that photon energy is proportional to $1/\lambda$. This suggested that as the nuclear charge increases, an electron in an L-shell feels a stronger attraction (remember Coulomb's Law) and thus loses more energy when it drops to the K-shell. Moseley's finding put to rest long-held beliefs that elements should be arranged in order of atomic weights. The linear relationship he found suggested instead that the properties of elements (such as the wavelength of K_α x-rays) are a function of atomic number. For example, his data showed that potassium was just ahead of argon in atomic number. In a modern periodic table, elements are placed horizontally in rows according to atomic number. The rows are arranged to yield columns of elements having similar properties (due to their similar outer shell configurations).

3.4 ➤   THE PERIODIC TABLE

The sketch below outlines some key regions of a modern periodic table.

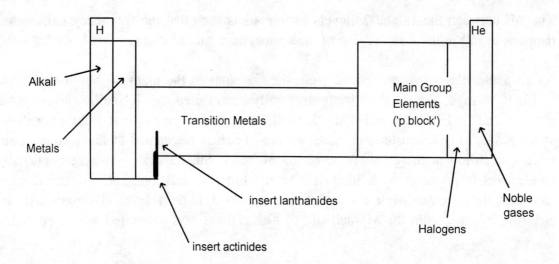

You should become familiar with these regions and the positions of the first 18 elements (**H to Ar**).

Besides the similarities in chemical properties among elements within a group, there are other important trends that are revealed by the periodic table. An example is the ionization energy or potential, which is a measure of the tendency of an atom in the gas phase to lose an electron and form a positive ion. Inspection of the table indicates that atoms with the lowest ionization potentials can be found on the left side, and in particular at the bottom left. Conversely, elements with high ionization potentials are found on the upper right side.

Electron affinity is a term associated with the tendency of an atom in the gas phase to capture an electron. Notice that electron affinities are negative for halogens such as chlorine and fluorine, indicating that energy is released upon electron capture by these elements. Electron affinities for alkali metals such as Na are positive, meaning that energy has to be imparted to make the negative ion of sodium.

These trends are consistent with observations about the chemical reactivity of elements. Sodium is likely to lose an electron (low ionization energy) whereas chlorine gains electrons readily (large negative electron affinity). Contacting sodium metal with chlorine gas is quite spectacular, with much heat being liberated with the production of sodium chloride.

Your periodic table also has electronegativities of the elements. Electronegativity is based on the following idea. Imagine a chemical bond between any two elements of the table. Electronegativity is the tendency of one of the atoms to pull electrons in the bond toward itself. The values range from about 0.9 for Rb to 4.1 for F. Notice the general correlation between electronegativity and ionization potential. In other words, elements which does not easily lose electrons tend to pull electrons toward them in chemical bonds, which is sensible.

4

CHEMICAL BONDING

4.1 ➤ TYPES OF CHEMICAL BONDS

All chemical bonds are electrostatic in nature; they are the result of electrostatic attraction between oppositely charged species. Primary chemical bonds, which are the strongest, include ionic, covalent and metallic bonds. Weaker secondary bonds are responsible for attractive forces between molecules. We will discuss secondary bonds last (Section 8).

The simplest case is ionic bonding, where negatively and positively charged ions interact with each other. An example is sodium chloride, which is composed of Na^+ and Cl^- ions. It is important to note that more than one sodium ion (actually, six) are attracted to each chloride ion and vice versa (see Chang, p.55). Ionic bonds are said to be non-directional (i.e., there is no single axis along which an ionic bond can be identified) because of this multi-ion interaction. The ions that make up ionic compounds are derived by reacting highly electronegative elements (e.g., halogens, Group 7A of the periodic table) with highly electropositive elements such as the alkali metals (Group 1A).

Covalent bonds are the result of sharing of a pair of electrons by two atoms. These bonds are typically formed from elements having rather small electronegativity differences. Covalent bonds are said to be directional because there is an axis along which the electron pair is shared. They form because each electron of the pair can be attracted to two nuclei simultaneously rather than one nucleus in the individual atoms. This statement is a bit too simplistic in that it cannot account for why helium is not diatomic. As you probably know, He has a filled $1s^2$ shell and you have learned that filled shells are stable. However, a much clearer picture comes from an application of rules from quantum theory which will be taken up in the next section.

A covalent bond drived from two identical atoms is said to be pure covalent, while one derived from dissimilar atoms is termed polar covalent. Dissimilar atoms have different electronegativities, meaning that one of the atoms will have a greater tendency to attract the electron pair toward itself. (An example is water.) Therefore, there will be an imbalance in the electron cloud distribution of the bond. The importance of this imbalance will be discussed in Section 8.

Metallic bonding is the result of sharing of electrons by many atoms simultaneously. This will be covered later (Section 7) once we have had a chance to explore a few more details.

4.2 ➤ AN IMPORTANT CONSEQUENCE OF QUANTUM MECHANICS: The Number of Orbitals is Conserved When Bonds are Made

Two atoms of hydrogen will spontaneously form diatomic hydrogen, H_2. This occurs simply because H_2 is more stable than H atoms. In H_2 we picture a single covalent bond holding the two atoms together. But how many orbitals were formed when the bond was made? Quantum mechanics makes the answer clear: two orbitals are formed. The two H atoms each have one atomic orbital (1s), and **the number of orbitals must be conserved** in the resulting molecule. There are two molecular orbitals (MO's) created from two atomic orbitals (AO's). Also, **the total energy of the orbitals must remain constant**.

What does all this mean? Let's consider the formation of H_2 from two H atoms in terms of energy. The atoms have some arbitrary but equal AO energy before reacting, as shown in Figure 1. We know that H_2 is more stable, so we expect an MO for H_2 to be at a lower energy compared with the atoms (shown in the Figure). This is called the bonding molecular orbital.

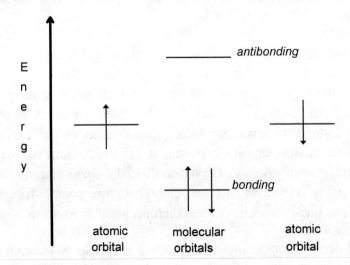

Figure 1. Orbital energy diagram for H atoms and diatomic hydrogen

However, there must be another MO. The only possibility is an orbital of an energy higher than that of the AO's by the same amount as the bonding orbital is lower. This is because the total energy must be conserved. This orbital is called an anti-bonding orbital. Where do we place the two electrons from the AO's which were used to form the covalent bond? Applying the rules discussed above, we are forced to place the two electrons in the bonding orbital as shown. The anti-bonding orbital is so named because electrons placed in it are at a higher energy than in the isolated atoms, and thus the molecule has no business existing (and won't).

An awareness of the presence of anti-bonding orbitals is essential to an understanding of important material properties. Without them we cannot explain, for example, the conductivity and luster of metals or the formation of semiconductor diodes.

Problem 4-1
Using the ideas developed in the last section, propose why He is always monatomic.

Problem 4-2
Also using these ideas, draw a molecular orbital energy diagram for Na_2, Na_3, and Na_4. Use only the 3s orbitals of Na - ignore the others. (We will discuss these diagrams in much more detail later in the course. If you understand the diagrams in a very general way, you are in a position to understand why sodium is a conductor of electricity.)

4.3 ➤ LEWIS DOT STRUCTURES

A Lewis dot structure is a short-hand notation describing an ion or molecule which shows all valence electrons. Some examples: a hydrogen atom is H·, diatomic hydrogen is H-H, sodium ion is simply Na+, and ammonia is

$$
\begin{array}{c}
\ddot{} \\
H{-}\overset{\displaystyle ..}{N}{-}H \\
| \\
H
\end{array}
$$

Notice that N is from group 5A and therefore has five valence electrons. Three are accounted for in the covelent bonds to the H's, and the other two remain on the N (the pair of dots, sometimes called a lone pair). The covalent bonds can be drawn as lines as shown above for ammonia and H_2, or they can be designated by electron pairs (dots) between the atoms. In order to determine if atoms in a molecule have a net charge, count all lone pairs and one electron per covalent bond and compare this total to the group number. If it is the same, there is no charge on the atom; if it is less than the group number by one, there is a single positive charge; if it is greater than the group number by one, there is a single negative charge. (If the electron count is less than the group # by 2, then there is a +2 charge, etc.) Convice yourself that the ammonium ion has a + charge (NH_4^+), and O with four lone pairs ($1s^2 2s^2 2p^6$) has a charge of 2-.

4.4 ➤ THE SHAPES OF MOLECULES

Consider the following facts:
- Antibodies can recognize antagonistic molecules (antigens) with remarkable specificity.
- Graphite, a form of pure carbon, is soft (the 'lead' in a pencil), whereas diamond (also carbon) is one of the hardest substances known.
- Molecules such as sucrose, saccharine and Nutrasweet are, literally, sweet.

Molecular geometry, which is ultimately dependent on bonding type, is responsible for each observation. For example, 'sweet' molecules interact with receptor sites in the tongue that bind only these molecules, activating a taste bud. The molecules must have a shape (geometry) which can bind in the molecular pockets of 'sweet' taste buds. Also of importance are secondary bonding forces (introduced in the next lecture) which ensure that the molecule is attracted to (and held to) its target. (Antibodies bind antigens in a similar fashion.)

The **v**alence **s**hell **e**lectron **p**air **r**epulsion (**VSEPR**) theory is used frequently to predict molecular geometry. This concept is not nearly as complicated as it sounds. It is based on the simple idea that valence electron <u>pairs</u> (in covalent bonds or as lone pairs) want to be as far away from each other as possible. For example, how would you distribute three e-pairs about a central atom (e.g., BF_3) using this idea? The answer is in a plane, with 120° bond angles. In the case of CH_4 or $:NH_3$, the predicted geometries are tetrahedral with 109.5° bond angles. Indeed, the bond angles in methane are 109°28'. In ammonia, the H-N-H angles are about 107°. The slight deviation from the 'pure' tetrahedral angle for ammonia is presumably because lone pair/covalent bond pair repulsions are greater than bond/bond repulsions, leading to a slight squeezing of the H-N-H angles in ammonia.

4.5 ➤ HYBRIDIZATION

The number of compounds that contain carbon vastly exceeds the number of compounds that do not. Inspection of the electron configuration for carbon, $1s^2 2s^2 2p^2$, suggests that carbon might use only its 2p electrons in forming covalent bonds (e.g.,

CH_2). However, carbon prefers to use all four of its n=2 electrons to form covalent bonds. Carbon is thus tetravalent, and can form a total of four covalent bonds with other atoms, including other carbons. These are not limited to single bonds; double and triple bonds are also possible. **Orbital hybridization** is a proposal which accounts for the bonding patterns of carbon. To be useful as a model, it must explain experimental facts. For example, the simplest hydrocarbon, methane (CH_4), is tetrahedral and all four C-H bond lengths are equal. Hybridization accounts for this. Hybridization is also useful in appreciating why the octet rule is violated by many molecules (e.g., AsF_5).

Hybridization literally means 'mixing.' Quantum mechanics allows orbitals to be mixed, at least mathematically. Again, the number of orbitals and total energy must be conserved. Let's develop these ideas with a focus on carbon.

The s and p orbitals of carbon can be hybridized to various extents depending upon the types of covalent bonds that are formed. The hybridization process itself requires energy, but this is more than compensated for by the greater number of bonds which can be formed. Consider the cases of methane. According to the hybridization model, the carbon atom can be viewed as employing its **one** s-orbital and all (**three**) available p-orbitals to form four new sp^3 orbitals; the resulting molecule has the shape of a tetrahedron (Chang, Fig. 10.8). In the structure of ethane, C_2H_6, both carbons of ethane are sp^3-hybridized, and the atoms surrounding each form the corners of a tetrahedron.

Now consider ethylene, $CH_2=CH_2$ (Chang, Figs. 10.14 - 10.16). Note that only two of the available p-orbitals have been hybridized, with the remaining one being perpendicular to the molecular plane. The "pure" or unhybridized p-orbitals form a new bond, a p-bond, through mutual "side-by-side" overlap. The degree of orbital overlap here is less than that of the C-C σ-bond, and consequently the former is weaker. This lower bond energy implies greater reactivity and in fact many important substances are prepared by reactions which open π-bonds.

4.6 ➤ DELOCALIZED BONDING AND RESONANCE

A covalent bond connotes sharing of a pair of electrons between two atoms, and it is a relatively easy matter to express this bond by a line (or lines in the case of double and triple bonds) between two atoms. We recognize that this line represents a bonding orbital in which the two electrons reside. However, let's think about why covalent bonds form in the first place from an electron's point of view. A valence electron in an atomic orbital is attracted to the atom's nucleus since opposite charges attract. When a covalent bond is formed using that valence electron, the electron is in a more diffuse molecular orbital and can now be attracted to <u>two</u> nuclei simultaneously. This simultaneous attraction lowers the energy of the electron, and this is why bonding MO's are lower in energy than AO's. Another way to say it is that the electrons in a molecular orbital are more delocalized (spread out in space) than in an atomic orbital, and this delocalization is responsible for the stability of covalent bonds. It follows, then, that if electrons have the opportunity to interact with more than two nuclei, they typically will do so. Thus, covalent bonds have a tendency to become even more delocalized. This is the essence of resonance.

One of the earliest examples of delocalized bonding is the carbonate ion, CO_3^{2-}, where x-ray studies proved that the bond lengths were equivalent. Linus Pauling argued that this result is explicable if one assumes that the π-bonds in carbonate ion are <u>delocalized</u> over more than two atoms. In other words, the electron pair of what our simple picture would predict to be one π-bond is actually shared equally among all three possibilities (shown in the middle of p. 354 of Chang).

The 'real' carbonate ion is a hybrid of these three *resonance forms* or *resonance structures*. Since each of these are of equal energy, they contribute equally to the *resonance hybrid*, which can be drawn as follows.

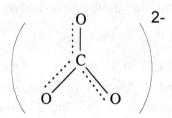

This notation is <u>awkward</u>, and we always prefer to write a Lewis structure for carbonate as any one of the three shown previously. However, we do so with the understanding that any of these does not represent the true structure. Note well: <u>the hybrid is a unique entity—it is not time-dependent, and thus does not "look" like any of the contributing structures at any split-second.</u>

In a nutshell, resonance is a manifestation of the fact that electrons, when possible, <u>like to be shared among more than two atoms</u>. In other words, electrons like to be <u>delocalized</u>.

4.6 ➤ VISUALIZING THE SOURCE OF RESONANCE

How can we tell when delocalized bonding is possible? Most of the time it is not too difficult, and the ability to pick out opportunities for delocalized bonding will assist greatly in the understanding of why certain chemical reactions occur and, if they do, at what rate. Any covalent bond is the result of overlap of orbitals. It should come as no surprise, then, that delocalized bonding is made possible by orbital overlap. Let's return to carbonate to understand this, drawing on our appreciation for the rules and predictions of VSEPR theory.

First consider any one of the resonance forms of the carbonate ion. VSEPR predicts that the ion will be planar, with 120° bond angles. This is because the carbon atom is sp^2-hybridized. The p orbitals which overlap to form the π-bond are perpendicular to the C-O σ-bond. Of major importance here is the fact that the p-orbital containing one of the three lone pairs on an adjacent O⁻ happens to be <u>parallel</u> to the lobes of the π-bond. This subtle but crucial fact provides the opportunity for delocalized bonding. Now imagine this lone pair moving toward the carbon, while the two electrons in the π-bond move to the O that was originally participating in the π-bond.

So now where is the π-bond? It has apparently 'moved' to another position. Or has it? It could equally well 'move back' to its original position. Alternatively, since there is yet another O with a lone pair which is parallel to the 'new' π-bond, one can generate yet another p-bond by the same mechanism. <u>Now</u> where is the π-bond? You might be tempted to think that the p-bond oscillates among these positions. Forget it—it's <u>wrong</u>. <u>The two electrons of the π-bond are always shared equally among the four atoms</u> (one C and three O's)—this is an example of <u>resonance</u>. The hybrid shown earlier is more stable than any of the three contributing structures because sharing of two electrons between more than two atoms provides stabilization energy for the molecule (sometimes called resonance energy).

Remember—delocalization is possible only because of overlap of orbitals. In other words, there must be a path over which electrons can travel in order to delocalize. That path in carbonate is provided by overlap of one of the filled p-orbitals on each of two O's with the π-bond. Resonance theory nicely explains why the bond lengths in benzene (C_6H_6) are equal. A common (but inaccurate) way to represent benzene is as follows:

It is inaccurate because the structure predicts distinguishable single and double bonds, which is not consistent with experiment (the bond lengths are equal). Let's look at this structure of benzene in more detail.

Note that if we were to, say, 'push' each pair of electrons in the π-bonds to the left, we would generate

This is still incorrect. The six π-orbitals making up the p-bonds are coplanar and provide the necessary overlap to allow the 'flow' of electron density to yield this structure. However, this overlap allows them to be shared equally among all six carbons, and energetically it is desirable to do so. Both resonance structures shown above contribute equally to the hybrid because they are of equal energy, and benzene is commonly represented by the structure below which attempts to reflect the delocalization of the π-electrons.

4.7 ➤ METALLIC BONDING

A perusal of the periodic table reveals that most of the elements are metals. The non-metals are found on the right side of the table, but even within the main group elements there are metals (e.g., Pb and Al). Metals have important properties such as electrical and thermal conductivity and reflectivity, and these properties are the direct result of the electronic structure of the constituent atoms.

Consider the alkali metals (Li, Na, K, etc.) first. We know these metals are very reactive, as is predicted by their positions in the periodic table. As an example, elemental sodium is a malleable solid, shiny when pure, and a decent electrical conductor. Since it has a valence of 1, it is tempting to think that sodium is diatomic. Indeed, sodium vapor contains Na_2 molecules, but solid sodium contains an ordered array of atoms, with each Na surrounded by eight other Na's. There are no obvious covalent bonds between a particular pair of atoms. The atoms of solid sodium, like in other metals, are held together by what are known as **metallic bonds**.

4.7a ➤ THE 'SEA-OF-ELECTRONS' MODEL

Imagine an array of Na atoms. Now remove the 3s electron from each, and imagine the electrons are free to wander through the array of the resulting Na$^+$ ions. No 3s electron is associated with any particular Na atom. This is a crude picture of metallic bonding. Each valence electron is thus shared by many Na's.

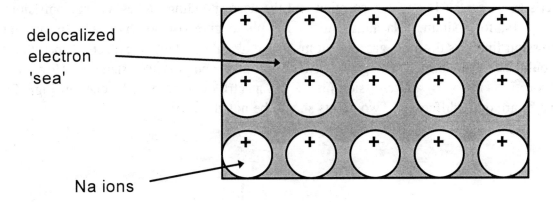

delocalized electron 'sea'

Na ions

This is an example of large-scale resonance. The 'electron sea' is responsible for conduction—since the valence electrons are not tied down in covalent bonds, they can move in an electric field.

The alkali metals are malleable and have rather low melting points because the individual metal-metal bonds of the 'electron sea' are weak. Many metals, however, have true covalent bonds between atoms in the solid along with a delocalized sea of electrons. This is typically the case with transition metals, where s electrons form the 'sea' and unpaired d electrons are used to form covalent bonds. Tungsten is very hard and high-melting for this reason.

4.7b ➤ ABSORPTION AND EMISSION SPECTRA OF METALS:
The Energy Band Picture

The interaction of materials with electromagnetic radiation is the most powerful method of probing atomic and electronic structure. Atoms or molecules may absorb light upon electron excitation from ground to higher energy states. A profile of absorption intensity as a function of wavelength or frequency (energy) constitutes an absorption spectrum. Excited atoms or molecules can emit light upon returning to the ground state; the latter phenomenon gives rise to an emission spectrum. Remember that the Bohr theory and the eventual development of quantum mechanics resulted from the desire to explain spectroscopic data. Even now, data from spectroscopic measurements are being used to refine theories of bonding. Some early spectroscopic experiments on gaseous metal atoms and their corresponding solids helped to clarify what happens when isolated atoms interact to form solids. The spectra can be interpreted with the aid of simple quantum theory.

Let's focus on the 1s → 3s absorption and the corresponding highest energy emission, 3s → 1s, for sodium. Na atoms in the gas phase give rise to sharp absorption and emission lines, as do most gaseous atoms and molecules. This is expected since there should be well-defined energies between ground and excited states. However, absorption (1s → 3s) and emission (3s → 1s) spectra of solid Na (sketched in Figs. 2a and b) are quite different. Two points should be noted. First,

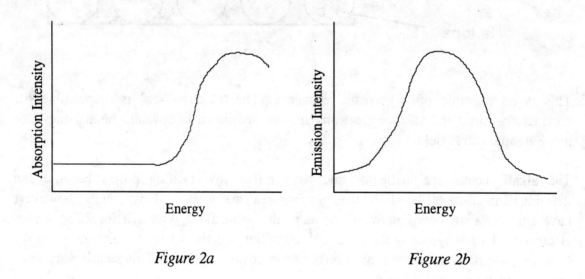

Figure 2a   Figure 2b

the spectra are broad. Second, the beginning of the absorption spectrum overlaps with the tail of the emission spectrum.

The absorption spectrum suggests that the 1s electrons are being promoted not to one excited state of a unique energy, but rather to a range of empty energy levels, as suggested by the breadth of the spectrum. Furthermore, the energy differences between the many empty states appear to be small, as suggested by the absence of well-defined peaks. Eventually no more states are available, and the absorption intensity drops. A possible energy level diagram consistent with the spectrum, showing a few of the electronic transitions, is:

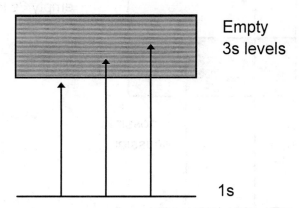

The 3s levels are all empty in this diagram, and only a few of the possible 1s → 3s transitions are shown.

Now look at the emission spectrum. To generate the spectrum, an electron from the 1s level is ejected by electron bombardment, and electrons from higher levels drop down (recall x-ray generation). As a reminder, we are looking at the 3s → 1s emission. A possible energy level diagram accounting for the emission spectrum is:

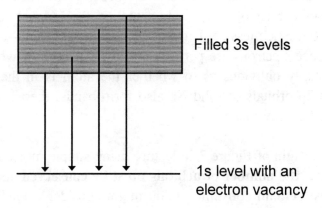

But the filled 3s levels depicted here <u>must be at different energies</u> than the empty ones giving rise to absorption. In fact, they must be at <u>lower</u> energies. For example, the lowest energy 3s → 1s emission event is far lower in energy than the initial absorption. Considering this and the fact that the spectra do overlap a bit (suggesting that the energies of the lowest energy absorptions are similar to those of the highest energy emissions) leads us to Figure 3.

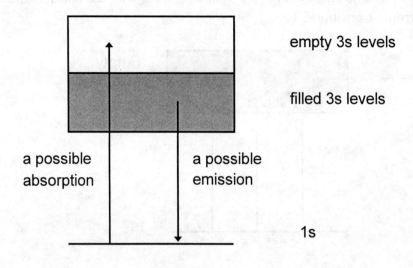

Figure 3

This diagram is quite consistent with the experimental results. The 3s levels can best be represented by an **energy band** that is <u>partially filled</u>. Absorption takes place between the 1s level and the empty levels of the band, and emission occurs from filled 3s band levels to an electron vacancy created in the 1s level. The energies between individual levels within the band must be small because the spectra appear as broad envelopes rather than a series of spikes. In actuality, the 1s level also broadens into a band in the solid,[1] but much less so than 3s. <u>This is a key point</u>—it means that in the solid state, the valence electrons are perturbed to a great degree, while the inner-most electrons are essentially oblivious as to whether the atom is in the gas phase or the solid. (The 2s and 2p orbitals of solid Na also form bands; the amount of broadening increases as the shell energy increases.)

We will explain the origin of Figure 3 in lecture using simple molecular orbital theory. Remember the rules: **the number of orbitals must be conserved** and **the total energy of the orbitals must remain constant**. (You might consider extending problem #2 to "Na_n", where n is a large number of sodium atoms. You should obtain 1s, 2s, 2p and 3s 'bands,' the last being <u>half-filled</u>.)

One final point: the molecular orbitals formed when many Na atoms condense to form a solid <u>extend throughout the solid</u>. This has extremely important implications for technology.

---

[1] The 1s level (and all others) must split into a band of energies, even if the energy differences are exceedingly small. Otherwise the collection of 1s electrons will violate the Pauli exclusion principle.

4.8 ➤ INTERMOLECULAR FORCES

4.8a Permanent and Instantaneous Dipoles

Hydrogen chloride, because of the electronegativity difference between H and Cl, has more electron density at the Cl atom. The molecule can therefore be written as

$$\overset{\delta+}{H}\text{--------}\overset{\delta-}{Cl}$$

There are partial charges (designated by $+\delta$ and $-\delta$) rather than full + and - charges because HCl is not completely ionic. The bond between H and Cl is known as a polar covalent bond, and is a <u>dipole</u>. It should not be surprising that the $+\delta$ end of an HCl molecule will be attracted to the $-\delta$ end of another, and this in fact is an example of an intermolecular attraction.

A dipole has dipole moment, μ, which is defined as the product of the magnitude of the charges and the distance separating the charges. Dipole moments can be measured experimentally, and are typically reported in Debyes.

It is important to note that in molecules containing more than two atoms it is necessary to sum the individual bond dipoles to obtain a net dipole moment for the molecule. For example, carbon tetrachloride, CCl_4, possesses polar C-Cl bonds but yet its dipole moment is zero. This is because vector addition of the individual bond dipoles in this symmetric molecule leads to a resultant of zero. Molecules with little or no dipole moment are termed <u>non-polar</u>.

Non-polar molecules, however, are continuously generating instantaneous dipoles. Electron clouds at a given instant may not be symmetric about an atom or molecule. There may be regions of higher and lower electron density, leading to instantaneous dipoles (Chang, Fig. 11.5). When one forms, it can induce a slight electron density fluctuation in a neighboring molecule (the $+\delta$ end of the first pulling electron density from the second toward it). The result is a weak attractive interaction. It quickly disappears, but is constantly being replaced by others forming nearby. Therefore, even non-polar molecules have attractive forces between them, although these are typically weaker than the forces between polar molecules. It should be noted that all molecules, including polar ones, generate instantaneous dipoles by this mechanism.

Intermolecular attractive forces are, as expected, far weaker than covalent bonds. For example, the bond dissociation energy for H_2 (corresponding to the reaction $H_2 \rightarrow$

2H•) is about 435 kJ/mol, while the energy of sublimation of solid H_2 is only about 0.5 kJ/mol.

4.8b-1 Scope and Importance of Intermolecular Forces.

All forces operating between molecules are electrostatic in nature; that is, they are attractive forces between oppositely charged species. The weakest of the intermolecular forces are known as Van der Waals or dispersion forces (ca. 0.5 - 10 kJ/ mol). These exist between all types of atoms and molecules, and are the result of instantaneous dipoles inducing the formation of neighboring dipoles as discussed earlier. Next in terms of strength are dipole-dipole forces (ca. 10-25 kJ/mol) which as is implied from the name operate between molecules possessing polar covalent bonds (such as HCl). The strongest intermolecular force is known as a hydrogen bond (ca. 25-40 kJ/mol), and it is really a special case of a dipole-dipole attraction.

Intermolecular forces are responsible for the formation of condensed phases (liquids and solids) of non-metallic atoms and molecules. In fact, strong intermolecular forces (specifically, hydrogen bonds) account for water's unusually high melting and boiling points considering its small size. Intermolecular forces serve to hold complex molecules into very precise shapes, such as DNA's double helix.

4.8b-2 Van der Waals or Dispersion Forces

Atoms and non-polar molecules have no dipole moment. Yet Ar can be liquified (it's boiling point is ≈ -185°C). Bromine is a liquid a room temperature, and iodine is a solid. The formation of condensed phases from these non-polar substances is the result of instantaneous dipoles as noted earlier. The more loosely held electrons are, the more polarizable they are; that is, the more easily they are coerced into electron density shifts by neighboring instantaneous dipoles. Electrons become more loosely held the farther they are from the nucleus, and so it is not surprising that large atoms are more polarizable than small ones. Also, larger atoms provide a larger surface area over which intermolecular forces can act. The net result is that dispersion forces are greatest for large, polarizable atoms and molecules. This is why iodine is a solid at room temperature whereas chlorine is a gas.

4.8b-3 Dipole-Dipole Forces and Hydrogen Bonds

Dipole-dipole forces are simple to understand—they result from attraction of opposite ends of a permanent dipole. These forces become particularly strong when the partial charges at the ends of the dipole are large. A classic example is water. Oxygen has a small radius and is electronegative, while H is very small and rather electropositive (the opposite of electronegative). The oxygen atom pulls electron density toward it, concentrating a partial negative charge on a fairly small atom. Likewise, the small H has a rather large $+\delta$ on it. This allows the O on one water molecule to approach the H on another rather closely, leading to a strong intermolecular attraction. Because on of the atoms participating in this attraction is H, the interaction is referred to as a hydrogen bond (Chang, Fig. 11.6).

Hydrogen bonding is primarily limited to molecules having H attached to N, O, or F. These molecules are electronegative and small, allowing for (1) large partial charges to develop, and (2) close approach of the molecules. By way of comparison, water boils at 100°C yet H_2S is a gas at room temperature. In the latter, dispersion and dipole forces exist, but S is too electropositive and large to make H_2S an effective hydrogen bond former.

Hydrogen bonds are frequently found between large biological molecules, and are responsible for holding the molecules in a specific pattern. See, for example, sketches of proteins (Chang, Figs. 26.13 and 26.15) and DNA (Chang, Fig. 26.23).

4.8c Solubility

A good rule of thumb is "like dissolves like." This suggests that non-polar (or relatively non-polar) molecules will dissolve in other non-polar (or relatively non-polar) ones, but reluctantly in polar molecules, especially those having intermolecular hydrogen bonds.

Water is a particularly good medium in which to dissolve ions. The Coulombic attraction between oppositely charged ions in, say, solid NaCl must be compensated by attractive forces between water and Na^+ and Cl^- for salt to dissolve. Water is effective because the partially positive H's can interact with Cl^-, and the partially negative oxygen can interact with Na^+. Non-polar molecules cannot provide such an opportunity, and experience tells us that salt does not dissolve in gasoline.

Question: Soaps are molecules containing a modestly long non-polar chain and a polar group at one end. How can these molecules clean grease (a non-polar substance) from clothes which are being soaked in a water 'solution' of the soap?

5

BONDING IN SOLIDS

5.1 ➤ INTRODUCTION

The distribution of electrons around an atom was discussed in Chapter 2 on atomic structure. Some electrons (s electrons) are distributed with spherical symmetry around the atomic nucleus, whereas other electrons have strong directional character (p, d, etc. electrons). These electron types are combined in atoms as discussed in Chapter 3 on the periodic table. The number and kinds of electrons determine the chemical behavior of atoms of different elements.

In solids, atoms are held together by bonding forces. These forces result from the exchange or sharing of electrons between atoms. There are the following types of bonds in solids: Ionic, metallic, covalent, and secondary. The first three types are discussed in this chapter, and secondary bonds, which are weaker than the other three, are discussed in Chapter 4. Covalent bonds in molecules are discussed in Chapter 4, together with more details on different kinds of covalent bonds.

5.2 ➤ LEWIS DOT SYMBOLS

In Chapter 3 on the periodic table the stability of noble gases such as neon, argon, krypton and xenon was described. The atoms of these elements have a completed outer shell of eight electrons, which is a stable form, and results in the low reactivity of these noble gases. In combinations of atoms to form molecules or solids, this same electronic structure of a complete shell of eight electrons also leads to stability. The American physical chemist, Gilbert Lewis, suggested that by sharing or exchanging electrons atoms can acquire a stable noble gas configuration.

There are two different ways to show this electron exchange or sharing to form a more stable configuration than individual atoms. One is to use a dot for each valence (outer) electron in an atom or ion; for stability each atom should have or share eight electrons. Consider for example the fluorine atom with an electronic structure $1s^2$, $2s^2$, $2p^5$. The superscripts designate the number of electrons in each state. Each fluorine atoms, therefore, has seven valence electrons. Two fluorine atoms can share electrons to form a stable molecule of F_2:

$$:\ddot{F}\cdot + \cdot\ddot{F}: = :\ddot{F}:\ddot{F}: \tag{1}$$

The total number of electrons in F_2 is the same as in two fluorine atoms, but by sharing two electrons in a <u>covalent bond</u>, the more stable F_2 molecule is formed. Many non-metals form stable gaseous diatomic molecules. Examples are hydrogen H_2, fluorine F_2, chlorine Cl_2, bromine Br_2, and iodine I_2. The latter two elements are liquid and solid respectively at room temperature, but form diatomic gases at temperatures somewhat above room temperature.

A second way to designate the stable sharing of electrons is to use a line to show a bond: F-F. This method is simpler, but if the reader has doubt about where electrons in a bond come from, the dot symbols can help.

5.3 ➤ IONIC BONDS

In solids atoms are bonded together in a three dimensional structure. Each atom is bonded to several other to atoms in the solid. However, it is easier to understand how atoms bond together if we start with just two atoms and see how they interact. In solids the individual bonds between atoms often are separate, so the solid is held together by many individual atom-atom bonds. In other solids this separation is not so simple. Nevertheless, we will start with atom-atom pairs, because we can learn much about bonding in solids this way.

Let us consider the two atoms that make up a common material, salt or sodium chloride. Sodium has the electronic structure $1s^2$, $2s^2$, $2p^6$, $3s^1$. Remember that the superscript gives the number of electrons in each state. The single 3s electron in sodium has a much higher energy than the other electrons, so it is easily given up to other atoms. Thus, sodium metal is highly reactive. The sodium ion resulting from a loss of this electron has a single positive charge.

The electronic structure of a chlorine atom is $1s^2$, $2s^2$, $2p^6$, $3s^2$, $3p^5$. Thus, the chlorine atom lacks one electron to fill the shell with quantum number 3 with eight electrons to form a stable ion. This chlorine ion has a single negative charge. In sodium chloride the 3s electrons from the sodium atoms have all been transferred to chlorine atoms to form chlorine ions. This electron exchange leads to a solid that contains equal numbers of positively charged sodium ions and negatively charged chlorine ions. Overall in the solid there are the same number of positive and negative charges, so the solid is electrically neutral; locally there are charged ions that result from the electron exchange. A positively charged ion such as sodium in sodium chloride ($Na+$) is called a <u>cation</u>, and a negatively charged ion such as chlorine (Cl^-) is called an <u>anion</u>. The Lewis dot symbols for these ions show the octet of electrons around them:

$$: \overset{\displaystyle ..}{\underset{\displaystyle ..}{Na}} :\,^+ \qquad : \overset{\displaystyle ..}{\underset{\displaystyle ..}{Cl}} :\,^- \tag{2}$$

To understand how the interaction of two oppositely charged ions leads to bonding, consider the forces on these two ions as they approach one another, as shown in Fig. 1. The two ions are attracted by an electrostatic force F given by

$$F = -\frac{a}{r^2} \tag{3}$$

in which r is the distance between the two ions and a is a constant. The negative sign denotes attraction. The dependence of F on r is shown as a dotted line in Fig. 1. It is the inverse square dependence on the distance between two electrically opposite charges. As the two ions approach more closely, the electron clouds around them begin to repel one another; this repulsion can be thought of as two atoms trying to occupy the same space. This repulsion force increases sharply as the distance of approach of the two ions becomes smaller, and is shown schematically in Fig. 1 as another dotted line. When the attractive and repulsive forces just balance, the net force on the ions is zero, and they are at their equilibrium or fixed separation. This separation is called the <u>bond length</u> between the two ions.

In solid sodium chloride the ions do not exist as individual pairs. Each sodium ion is surrounded by six chloride ions, which are called the nearest neighbors to the sodium, and each chloride ion also is surrounded by six sodium ions. (See Chapter 6 for more information on the structure of sodium chloride.) Thus, the actual electrical forces in a particular ion are not pair-like, but are closer to a spherically symmetrical electrical field. Nevertheless, the same principle of balanced forces, as for the pair, determines the interatomic distances in the real solid. There is a balance of forces resulting from electrostatic attraction and repulsion of electron clouds.

Atoms have different sizes depending upon their number of electrons and position in the periodic table, as described in Chapter 3. In general, atoms with more electrons are larger, although there are exceptions. Ions also have different sizes. When an atom loses an electron there is a net positive charge on the nucleus, and the electrons are pulled into the nucleus by electrostatic attraction. Thus, in solids cations such as sodium are smaller than a neutral atom such as neon with the same number of electrons. In the same way, anions are larger than atoms with the same number of electrons, because anions have an extra electron and a net negative charge, which leads to mutual repulsion of the electrons. Some ion sizes are given in Table 5-1, where these trends in size can be seen.

Table 5-1

Ion Sizes, in nm		
H^+ small	Mg^{++} .07	F^- .13
Li^+ .07	Ca^{++} .10	Cl^- .18
Na^+ .10		Br^- .20
K^+ .13		I^- .22
$O^=$ .14	Si^{+4} .004	$CO_3^=$ .13 OH^- .13
$S^=$ .18		NO_3^- .12 NH_4^+ .15

5.4 ➤ METALLIC BONDS

In a solid metal the individual atoms give up their valence electrons to form <u>free electrons</u> that are not associated with a particular atom but can move freely throughout the whole piece of metal. This free motion of electrons is what gives metals their high electrical conductivity. When a metal is subjected to an electrical potential, the free electrons move rapidly to give a high current.

As an example, consider solid aluminum. The electronic configuration of aluminum is $1s^2 2s^2 2p^6 3s^2 3p^1$. Thus, each aluminum atom gives up its three valence electrons ($3s^2$ plus $3p^1$) to the sea of free electrons. The remaining electrons stay tightly bound to the aluminum atoms and do not participate in the electrical conductivity.

Sample Problem:
Calculate the number of free electrons per unit volume in aluminum.

Solution:

The density ρ of aluminum is 2.70 gms/cm^3 and its molecular weight m is 27.0 grams/mole. Then the number of atoms per unit volume of aluminum is:

$$\frac{\rho A}{m} = \frac{2.70(6.02)(10)^{23}}{27.0} = 6.02(10)^{22} \; atoms \, / \, cm^3 \qquad \textbf{(4)}$$

in which A is the number of atoms per mole. Since each atom gives three free electrons, the number of free electrons per unit volume is $18.06(10)^{22}$ electrons/cm^3.

It is not easy to understand why this structure of free electrons and positively charged "core" atoms leads to forces that hold the atoms together. This understanding is beyond the scope of this course, so for now the reader should simply accept the stability of the metallic structure of free electrons and charged core ions.

The "electron in a box" solution of the wave equation (Section 2.9) is a good approximation to the behavior of a free electron in a metal. In this approximation the electrostatic repulsion between the free electrons, and the attraction between these electrons and the ion cores, is neglected. It is very surprising that these strong electrostatic forces do not influence electronic behavior more. This result shows that electrons sometimes behave in ways we do not expect from traditional electrostatics.

5.5 ➤ COVALENT BONDS

In covalent bonds electrons are shared between atoms. In the example in Eq. 1 of the F_2 molecule the two fluorine atoms are held together by the shared pair of electrons. Each fluorine atom supplies one electron to the bond, and the remaining electrons (non-bonding electrons) are localized on a particular atom.

In the sense of electron distributions discussed in Section 2.10, the electron clouds of the bonding electrons from the two atoms overlap. One might expect this overlap to result in an electrostatic repulsion, but in a bond the electron sharing actually reduces the total energy and stabilizes the bond.

An example of a covalently bonded solid is silicon. In silicon each silicon atom is surrounded by four other silicon atoms, as shown in Fig. 2. This structure is called tetrahedral bonding, because the four surrounding silicon atoms are at the corners of a tetrahedron, as shown in the figure. A tetrahedron is a symmetrical solid with four equal faces and four corners. Each silicon atom sits in the middle of a tetrahedron surrounded by four other silicon atoms on the corners.

The electronic structure of silicon is $1s^2 2s^2 2p^6 3s^2 3p^2$. Two bonding electrons are involved in each silicon-silicon bond. Around each silicon there are four bonds, so all the n=3 electrons take part in bonding. The possible electron states with n=3 are 2s states and 6p states. Since all the bonds are equivalent, they each have some s and some p character. These <u>hybrid orbitals</u> are described in more detail in Chapter 4.

5.6 ➤ ELECTRONEGATIVITY

The concept of electronegativity helps us to understand the tendency of an atom to attract the electron pair forming a covalent bond. The greater the electronegativity of an atom, the greater is its tendency to attract the electrons in a covalent bond. The American chemist, Linus Pauling, assigned an arbitrary scale to electronegativities from zero for the least to 4.0 to the most electronegative atom (fluorine). These electronegativities are given in Table 2 for a number of atoms. As the atomic number increases in a particular row of atoms in the periodic table, the electronegativity increases. There is also a small decrease in electronegativity as the atomic number increases in a column in the table.

Electronegativities help in understanding of bond strengths (see Chapter 8). The greater the difference in electronegativities between two atoms in a bond, the greater is the bond strength.

Electronegativities also show that the electron pair in a covalent bond shifts slightly to the more electronegative atom. For example, in the silicon-oxygen bond the electron pair is shifted slightly to the oxygen atom because it is more electronegative.

6

CRYSTAL STRUCTURE

6.1 ➤ INTRODUCTION

A solid can be defined as a material that does not flow when subjected to a force, in contrast to a liquid that does flow. If the atoms in a solid have a repeating ordered structure over many atom distances, the solid is called a crystal. Atoms in condensed phases (solids or liquids) have short range order; that is, the atoms and their close neighbors have a regular arrangement over a few atom distances. In a crystal this order extends over many atom distances; in a single crystal this order extends throughout the whole crystal. This long-range order means that planes of atoms in the single crystal extend from one crystal face to another with the same structure and orientation. A solid whose structure (atomic arrangement) has no long-range order is called amorphous or glassy. The atoms in such a solid or a liquid still have local (short-range) order, and no regular repeating structure over more than a few atom distances.

In this chapter, the structures of some simple crystal structures of elements are illustrated, and dimensional relations between atom sizes and repeating units in the crystal are derived. Then methods for defining directions and planes in the crystal lattice are described, and a few more complicated structures of compounds are shown. These structures lead to the concept of coordination number of an atom. Finally, the experimental method of X-ray diffraction for studying crystal structure is briefly described.

6.2 ➤ CUBIC CRYSTAL STRUCTURES

To illustrate the structures of crystals, atoms are considered to be spheres with a definite radius. From the discussions in Chapters 2 and 4 we know that the electrons around an atomic nucleus are spread out in a "cloud," although there is a most probable distance of each electron from the nucleus. In spite of this atomic structure, the "hard sphere" model for crystal structure is an accurate way to describe the structure. This description succeeds because the important factor in crystal structure is the arrangement of atoms into fixed groups, especially planes. The nuclei of the atoms have fixed positions that can define the crystal structure.

This hard-sphere model is especially useful for describing the structures of pure metals. In most of these structures, the atoms behave as if they were hard spheres, and pack together in a regular close-packing of the spheres. The most common metallic crystal structure is face centered cubic, shown in Fig. 1. In this figure the structure is drawn in two different ways. In both cases, the starting point is a cube. In Fig. 1a the atoms are represented as dots where their centers reside. In the face-centered structure there are atom centers at each of the eight corners of the cube and at the centers of each of the six faces of the cube. In Fig. 1b the atoms are shown as spheres that pack tightly in the structure. The atoms in the face-centered cubic structure are packed as tightly as atoms can be, as shown in the next section.

The cubes in Fig. 1 are called a unit cell of the structure. A *unit cell* is defined as a small repeating volume of a crystal lattice used to describe the crystal structure. The unit cell is chosen for convenience in visualizing the structure of the crystal. It is not necessarily the smallest repeating unit of the crystal structure. The crystal lattice is generated by unit displacements of the unit cell. The cube edge is called the *lattice parameter* of the unit cell or of the crystal structure.

Let us look at the atoms in the front face of the face-centered cubic lattice, as shown in Fig. 2. The atoms touch along the face diagonal in this structure, so this diagonal is called a close-packed direction. The face arrangement allows us to calculate a relation between the radius **R** of an atom in the structure and the lattice parameter (cube edge) **a**, as shown in Fig. 2. The length **d** of the face diagonal equals **4R**, and also $d^2 = 2a^2$. Therefore,

$$a = 2\sqrt{2}R \quad FCC \tag{1}$$

in the face-centered cubic structure.

Some metals that have face-centered cubic structure (see appendix) are aluminum, copper, gold, lead, nickel, platinum, and iron above $910^{\circ}C$.

Another cubic structure is body-centered cubic, as shown in Fig. 3a and b. Here the eight atom centers are on the corners of the unit cell cube, and one is at the center of the cube. The atoms touch along the diagonal of the cube, so that the relation between atomic radius and lattice parameter is

$$a = 4\frac{R}{\sqrt{3}} \quad BCC \tag{2}$$

Some metals that have the BCC structure are chromium, vanadium, niobium, tungsten, molybdenum, the alkali metals, and iron below $910^{\circ}C$.

One can also imagine a cubic structure with atoms at the eight corners of the cube only, which is called the simple cubic structure. However, it is rarely found in nature. Another important cubic structure is that of diamond, shown in Fig. 4. This structure is more open than the others we have considered, has more atoms in the unit cell, and is harder to visualize. It is best thought of as a more open face-centered cubic structure, with additional atoms in the biggest holes in the resulting structure. Carbon (diamond), silicon, germanium, and one form of tin all have the diamond structure. Because of the importance of silicon in the electronics industry, this structure is highly important.

6.3 ➤ DENSITY AND PACKING FACTOR

These quantities can be calculated from any unit cell from a knowledge of how many atoms are in the cell. To find the number of atoms in the unit cell, one must visualize how much of each atom is in the cell. For the face-centered cubic structure, there are eight atoms at the cube corners, and one-eighth of each of these atoms is in the cell giving one equivalent atom. These atoms are shared by eight other unit cells. Each face atom is half in the unit cell, and there are six of them, so the face atoms give three equivalent atoms, making a total of four atoms in the unit cell.

$$
\begin{array}{ll}
\text{corner atoms} & 8 \times \frac{1}{8} = 1 \\
\text{face atoms} & 6 \times \frac{1}{2} = \underline{3} \\
& \qquad\qquad 4 \text{ total}
\end{array}
$$

In the body-centered cubic there are two atoms in the unit cell, one from the corners and one in the center of the cell.

The formula to calculate the density P of a material from its unit cell contents and dimensions is:

$$P = \frac{nA}{VN} \tag{3}$$

for a material with one kind of atoms, in which **n** is the number of atoms in the unit cell, **A** is their atomic weight, **V** is the volume of the unit cell, and **N** is Avogadro's number, $6.02(10)^{23}$ molecules/mole.

Sample Problem:

Calculate the density of aluminum from its atomic radius of 0.143 nm and atomic weight of 27.0 grams/mole. The structure of aluminum is FCC, so its lattice parameter is $2\sqrt{2}R = .404$ nm. Then v = .0660 nm^3 and

$$P = \frac{4 \times 27.0}{.066 \times 10^{21} \times 6.02 \times 10^{23}} = 2.72 \frac{gms}{cm^3}$$

The measured value is 2.71 gms/cm^3.

The packing factor of a structure can be found from the ratio of the volume of the atoms in the unit cell to the volume of the unit cell itself. In the face-centered cubic structure, the volume of the atoms is $4 \times 4\pi R^3/3$ and the volume of the unit cell is $a^3 = 16\sqrt{2}R^3$, from Eq. 1. The packing factor is:

$$\frac{16\pi R^3/3}{16\sqrt{2}R^3} = \frac{\pi}{3\sqrt{2}} = 0.7406$$

For the body-centered unit cell, the volume of the unit cell is $a^3 = 64R^3/3\sqrt{3}$, so the packing factor is:

$$\frac{8\pi R^3/3}{64R^3/3\sqrt{3}} = \frac{\pi\sqrt{3}}{8} = 0.6802$$

6.4 ➤ LATTICE DIRECTIONS AND PLANES

In describing crystal structures is it useful to have a method of designating directions in the unit cell or crystal lattice, and planes in the lattice. A direction in the crystal lattice can be thought of as a vector from the origin of a coordinate system to a point in the lattice. It is most convenient to place the origin of a coordinate system at a corner of the unit cell, as shown in Fig. 5. The lattice direction is just the vector from this corner to another point in the lattice. For example, a direction of a face diagonal in the **y-z** plane is [0,1,1], as shown in the figure. The cell diagonal is then [1,1,1]. A [1,$\bar{2}$,0] direction is also shown in the figure. The negative direction along the y-axis is designated with a bar over the 2. Individual directions are designated by square brackets [011] and families of directions by diamond brackets. Thus, the family of <100> directions includes the directions [100], [010], [001], [100], [010], and [001].

To designate a plane in a crystal, consider the equation of a plane:

$$\frac{x}{a} + \frac{y}{b} + \frac{z}{c} = 1 \tag{7}$$

in which **a**, **b**, and **c** are the intersections of the plane with the axes **x**, **y**, and **z**, respectively. The plane can, therefore, be designated by the reciprocals of these intersections, which are called the *Miller Indices* **h**, **k**, and **l** of the plane:

$$h = \frac{1}{a}, \quad k = \frac{1}{b}, \quad l = \frac{1}{c} \tag{8}$$

An example is given in Fig. 6a of a (111) plane. The intersections of the plane with the three axes are a=1, b=1, and c=1, so h=k=l=1. Also, in Fig. 6b is a (213) plane. Here the intersections are $\frac{1}{2}$, 1, and $\frac{1}{3}$, so the Miller Indices are 2, 1, and 3. Miller Indices are always integer values. Thus, for intersections $\frac{2}{3}$, $\frac{2}{3}$, $\frac{1}{2}$, for example, the Miller Indices are (226) or more simply (113).

If the plane does not intersect an axis, the intersection is infinite, and its reciprocal is zero. Then, the Miller index of this plane for this axis is zero. Examples are in Fig. 7. In Fig. 7a the plane intersects only the x-axis at $\frac{1}{2}$, so its Miller indices are 2, 0, 0. In Fig. 7b the plane intersects the **y** and **z** axes at one, and does not intersect the x-axis, so its indices are 0, 1, 1.

Individual planes are designated with parentheses, for example (110). Families of planes are designated with curly brackets, for example, {100}. These planes include all possible ones: (100), (010), (001),($\bar{1}$00),(0$\bar{1}$0), and (00$\bar{1}$) six different individual planes. The {111} family includes (111),($\bar{1}$11), (1$\bar{1}$1),(11$\bar{1}$),($\bar{1}\bar{1}$1),($\bar{1}$1$\bar{1}$),(1$\bar{1}\bar{1}$), and ($\bar{1}\bar{1}\bar{1}$) planes, or eight total.

6.5 ➤ STRUCTURES OF COMPOUNDS

Structures of compounds such as MgO are more complicated than those of crystals containing only one kind of atom. We will describe two of the most common compound structures; more compound structures will be discussed in Chapter 12. Many solids have the same crystal structure as common salt, NaCl, including most alkali halides, divalent oxides and sulfides such as CaO, MgO, FeO, MnS, FeS, and PbS, carbides (ThC) and nitrides (ZrN). This sodium chloride structure is cubic and is shown in Fig. 8. The unit cell can have either sodium or chloride ions at the corners of the cube; in Fig. 8 the chloride ions are on the corners. In addition, chloride ions are at the face centers, so that the structure of the chloride ions by themselves is face-centered cubic. Sodium ions are between the corner chlorides; the sodium ions are on a face-centered cubic lattice. If all the ions are considered, they fall on a simple cubic lattice with a smaller lattice parameter than the face-centered lattices.

Another compound structure of cubic cesium chloride is shown in Fig. 9. In this structure one ionic type, for example chloride, is at the cube corners and the other at the center of the cube. Thus, the structure of the individual ions is simple cubic, whereas that of all the ions is body-centered cubic.

The unit cell dimensions of these crystals are related to the sizes of the ions. In the sodium chloride structure the ions are touching along the cube edges, so the lattice parameter **a** equals the sum:

$$a = 2r_1 + 2r_2 \qquad (9)$$

in which r_1 is the radius of the chloride ion and r_2 is the radius of the sodium ion. In the cesium chloride structure the ions are touching along the cube diagonal, just as in the body-centered cubic structure with all the same atoms. Therefore, the cube diagonal

$$d = \sqrt{3}a = 2r_1 + 2r_2 \qquad (10)$$

in which **a** is the lattice parameter and r_1 and r_2 are the ionic radii.

6.6 ➤ COORDINATION NUMBER

The number of atoms or ions that touch a central atom or ion in a crystal structure is called the coordination number of the central atom. In the body-centered cubic structure, the atom in the center of the cube touches eight other atoms, the ones at the cube corners, so its coordination number is eight. Since all the other atoms in the structure have the same surroundings, because of the repeatability of the structure, all the atoms have a coordination number of eight. Similarly in the cesium chloride structure the coordination number of both kinds of ions is eight.

In the close-packed structure of the face-centered cubic lattice, each atom is coordinated to twelve other atoms. This result can be visualized by considering an atom at a face center. It touches four other atoms in the facial plane (100). It also touches four atoms in the (020) plane and in the (002) planes, for a total coordination number of twelve. Thus, the closest packing of a structure with atoms of all the same size is achieved with a coordination number of twelve.

The coordination numbers of ions or atoms in structures is related to the bond type. On the diamond structure (Fig. 4), the coordination number of atoms is four, even though the atoms are all of the same size. Instead of the atoms packing tightly like non-interacting balls, the structure is determined by the strong directional covalent bonds between atoms such as carbon, silicon, and germanium. The diamond structure is more open than the close-packed face-centered cubic structure.

The coordination numbers of ions in ionically-bonded structures such as NaCl or CsCl are related to the ionic radii of the ions. In the cesium chloride structure each ion is surrounded by eight oppositely charge ions, and the ratio of ionic radii is 1.65(cesium)/1.81(chloride)=.912. If the radius ratio is smaller, a lower coordination number is favored in the structure. For NaCl, the ratio is .98(sodium)/1.81(chloride) =.541. The "ideal" radius ratio for a particular coordination number can be calculated from a structure of closest packing for that radius ratio. These ideal ratios are given in Table 1. One would expect the coordination number to be that for the next ideal ratio below the actual one. Thus, the ratio of .91 for cesium and chloride ions gives a coordination number of eight, and of .54 for sodium and chloride ions a coordination number of six. In real crystals, these ionic ratios are approximations; as discussed in the introduction to this chapter, the model of hard spheres is only an approach to the real structure of ions and atoms. Nevertheless, the structures of many ionic compounds can be deduced from radius ratios.

6.7 ➤ X-RAY DIFFRACTION

The determination of atomic crystal structures resulted from the discovery of X-ray diffraction. The wave-lengths of certain X-rays are close to the dimensions in unit cells of crystals. For example, the wave length of X-rays emitted from copper is about 0.15nm, which is similar to the distances between atoms in many crystals. Thus, X-rays can probe the structures of these crystals.

X-rays are electromagnetic waves just like light or radio waves. Therefore, they interact with the regular planes of atoms in a crystal just as light waves do with the regular grooves in a diffraction grating. The result is a set of sharp lines of maximum diffracted intensity when an X-ray beam reflects from the surface of a crystal. An experimental arrangement for studying X-ray diffraction from a powder of crystals is shown in Fig.

10. A beam of X-rays all of the same wave length reflects from the powder, and the intensity of the reflected beam is measured with a special detector as a function of the angle of incidence, shown in the figure. A typical diffraction pattern is shown in Fig. 11. The angles at which the X-ray intensity is a maximum are related to the spacing d between planes in the crystal by the Bragg equation:

$$\sin\theta = \frac{n\lambda}{2d} \qquad (11)$$

in which λ is the wave length of the X-rays, and **n** is an integer that gives the order of diffraction.

Since each family of planes in the crystal has a particular spacing, each diffraction angle θ (maximum in intensity) corresponds to a different set of lattice planes. Therefore, every crystalline material has its own "diffraction pattern" of sequence of angles θ that can be used to identify the material and its structure. The analysis of the phases present in an unknown material is the most important use of X-ray diffraction for chemists and materials engineers.

The crystal structures of a vast number of chemical compounds have been determined, and are tabulated in special references. Today the science of crystal structure determination continues to be important for highly complex crystals such as proteins and for new materials such as the oxide superconductors. Determination of these structures requires highly refined and specialized techniques; nevertheless, they are all based in the Bragg Law of Diffraction.

Table 6-1
Coordination Numbers and Radius Ratios of Ions
for Different Crystal Structures

Coordination Number	Radius Ratio	Examples of Structures
12	1	FCC
8	.732	BCC, CsCl
6	.414	NaCl
4	.225	SiO_2
3	.155	

7

EQUILIBRIUM OF CHEMICAL REACTIONS

7.1 ➤ INTRODUCTION

A simple definition of a closed system in equilibrium is that its chemical composition, properties, and structure do not change in experimental times. In this definition the time period that one must wait for changes to occur depends on the particular condition of interest, and so is somewhat vague. In practice this vagueness is not a problem, and conditions of equilibrium can be defined quite reproducibly for specific conditions.

It is possible to identify three types of equilibrium: mechanical, thermal, and chemical. For mechanical equilibrium there is no net unbalanced force in a system, for thermal equilibrium the system is at constant and uniform temperature, and for chemical equilibrium the system has constant composition and structure. In this chapter the equilibrium of chemical reactions is explored.

7.2 ➤ GASEOUS REACTIONS

Consider the reaction of carbon dioxide and hydrogen to form carbon monoxide and water, at a temperature above 100°C, so that all components are gases:

$$CO_2 + H_2 = CO + H_2O \qquad\qquad (1)$$
$$\quad A \qquad B \qquad C \qquad D$$

If these gases are mixed in arbitrary amounts in a closed container, they will react with time until they reach some final state of the mixture. This final state that does not change with time is called the equilibrium state of the reaction. In the special case of reaction 1 the total pressure in a fixed volume remains constant. The amount (concentration) of each gas can be described in terms of its *partial pressure*, as described in Chapter 5 in Chang. Let N_i be the number of moles of gas i in the container. Then from the ideal gas law (e.g., Chapter 5) the partial pressure of gas i is

$$P_i = \frac{N_i RT}{V} \tag{2}$$

in which R is the gas constant, T the temperature, and V the total volume of gas. If the reaction is studied at constant temperature, then the partial pressure is proportional to the concentration N_i/V in moles per unit volume of gas i.

The units of partial pressure are pressure units, for example atmospheres. The total pressure P_T is then the sum of all the partial pressures in the mixture; for reaction 1

$$P_T = P_A + P_B + P_C + P_D \tag{3}$$

In reaction 1 as hydrogen and carbon dioxide react they form carbon monoxide and water vapor until a final or equilibrium ratio is reached. Experiments show that the final ratio is given by an equilibrium constant K:

$$K = \frac{P_C \times P_D}{P_A \times P_B} \tag{4}$$

This equation says that for any initial amounts of the gases that are mixed together, the equilibrium mixture will obey Equation 4.

Sample Problem:
At 600°C the equilibrium constant K for reaction 1 is about .39 (unitless). If a mixture of carbon dioxide (partial pressure .4 atm) and hydrogen (partial pressure .6 atm) is held in a closed volume to equilibrium at this temperature, what is the final composition of the mixture?

In addition to Equation 4, the reaction itself imposes a constraint on the amount of the products produced. Let the equilibrium partial pressure of carbon monoxide be P_C. The goal is to solve for P_C, from which the other equilibrium partial pressures can be found. The sum of the partial pressure of carbon dioxide plus that of carbon monoxide always equals the initial pressure of CO_2 or 0.4 whether at equilibrium or not, because of reaction 1. Thus, at equilibrium $P_A + P_C = .4$ atm., and also $P_B + P_D = .6$ atms. Furthermore, for each mole of carbon monoxide produced, one mole of water is produced, so $P_C = P_D$. Putting these relations in Equation 4 gives

$$K = \frac{P_C \times P_C}{(.4 - P_C) \times (.6 - P_C)} \tag{5}$$

This is a quadratic equation in P_C whose solution is $P_C = .186$ atm. Thus at equilibrium the partial pressures of the various gases are $P_A = .214$ atm., $P_B = .414$ atm., $P_C = .186$ atm., $= P_D$.

Another commercially important gaseous reaction is the production of ammonia from nitrogen and hydrogen (the Haber process):

$$N_2 + 3H_2 = 2NH_3 \tag{6}$$
$$A \quad\quad B \quad\quad C$$

In this case, the total gas pressure in a closed vessel of constant volume changes, because two moles of ammonia are made from four moles of reactants. The equilibrium constant for this reaction is

$$K = \frac{P_C^2}{P_A P_B^3} \tag{7}$$

showing that the partial pressure of each reactant and product is raised to the power of the number of its molecules involved in the reaction. In the general case of gaseous reactants A, B, C - - and products E, F, G - - each with a coefficient (lower case letter):

$$aA + bB + cC\text{---} = eE + fF +\text{»---} \tag{8}$$

the equilibrium constant is

$$K = \frac{P_E^e P_F^f P_G^g \text{ --- }}{P_A^a P_B^b P_C^c \text{ --- }} \tag{9}$$

The direction a reaction goes can be found from the *reaction quotient* Q. Thus for reaction 1, the reaction quotient contains the partial pressures (lower case p's) at the beginning of the reaction:

$$Q = \frac{p_C \times p_D}{p_A \times p_B} \tag{10}$$

If $Q < K$, the reaction will go from left to right, and if $Q > K$, it will go from right to left. For example, consider the following gaseous mixture at 600°C. $P_A = .1$ atm., $P_B = .2$

atm., $P_C = .6$ atm., $P_D = .3$ atm. Then $Q = .6 \times .3/.1 \times .2 = 9$, and the reaction will go from right to left to form CO_2 and H_2.

In the systems considered so far, all the reactants and products were gases, so the reaction was *homogeneous*. In a system containing gases and a solid or liquid phase, the reaction is said to be *heterogeneous*. As an example, consider the reaction of solid carbon (graphite) with carbon dioxide to form carbon monoxide:

$$C + CO_2 = 2CO \qquad (11)$$
$$\text{A} \qquad \text{B} \qquad \text{C}$$

The equilibrium constant for this reaction is

$$K = \frac{P_B}{P_C^2} \qquad (12)$$

The carbon is considered to be at unit concentration as long as it is pure, so it does not have to be included in the expression for the equilibrium constant.

Another example of a heterogeneous reaction is the decomposition of a carbonate:

$$CaCO_3(s) = CaO(s) + Co_2(g) \qquad (13)$$
$$\text{A} \qquad\quad \text{B} \qquad\quad \text{C}$$

where (s) means solid phase and (g) gas. The equilibrium constant is

$$K = P_C \qquad (14)$$

so that at equilibrium the pressure of carbon dioxide is always the same at a particular temperature. This situation is similar to the vapor pressure of a pure compound, which is also a particular pressure at a particular temperature.

Sample Problem:

At 900°C the equilibrium constant for reaction 13 (Equation 14) is 1.04 atm. Thus, if some calcium carbonate is heated in a closed vessel, the pressure of carbon dioxide rises to 1.04 atm., where the decomposition of calcium carbonate stops. This result occurs no matter how much $CaCO_3$ and CaO solid phases are present, as long as they are pure and separate. If calcium carbonate is heated in an open vessel, it will continue to decompose until it is all converted to CaO, because the vapor pressure of CO_2 never reaches 1.04 atm.

7.4 ➤ LE CHATELIER'S PRINCIPLE

If a chemical reaction is at equilibrium, it can be shifted by various kinds of changes, for example, by changing the temperature or the amounts of reactants and products. The shift goes in the direction of minimizing the influence of the change. This process leads to a general statement called *Le Chatelier's Principle:*

> *If a system at equilibrium is subject to a change, the system will shift so as to counteract the effect of the change.*

As an example of the application of Le Chatelier's Principle, consider the equilibrium of reaction (1) at 600°C in the sample problem above. Let us now add some carbon dioxide to the reaction mixture. This action will cause the reaction to shift to the right, increasing the amount of CO and H_2O, and reducing the amount of H_2 in the mixture, to preserve the ratio of the equilibrium constant (Equation 4). Conversely, if some water vapor is added to the reaction mixture, the reaction will shift to the left. If the temperature is changed, the reaction will shift to counteract this change. For reaction 1 an increase in temperature causes an increase in the equilibrium constant; thus increasing the temperature of a mixture of reaction 1 at equilibrium will shift the reaction to the right, forming more CO and H_2O.

7.5 ➤ ACIDS AND BASES

Certain aqueous solutions have a set of properties that are characteristic, and these solutions are called acids. These properties are:

1. A sour taste. Vinegar and lemon juice contain organic acids, acetic ($C_2H_8O_2H$) and citric ($C_6H_8O_7$) respectively, that make them sour.
2. Reaction with many metals, such as zinc and magnesium, to form hydrogen gas, which forms bubbles.
3. Reaction with carbonates to form carbon dioxide.

The chemical species in water that gives rise to these properties is the hydrogen ion, H^+. This ion is a bare proton, without any electrons, so it is extremely reactive. In water it is always associated with a water molecule as the *hydronium* ion H_3O^+; it may actually be associated with more than one water molecule. In formulas either H^+ or H_3O^+ will be used.

Another group of aqueous solutions also have a certain set of properties; these solutions are called bases, and the properties are:

1. A bitter taste.
2. A slippery feel.
3. Reactions with acids to form salts and water.

The chemical species in water that gives rise to these properties is the hydroxyl ion, OH^-. Thus, in water acids form H^+ ions and bases OH^- ions. Examples of neutralization reactions of acids and bases are:

$$HCl + NaOH = NaCl + H_2O \qquad (15)$$
$$\text{acid} \quad \text{base} \qquad\quad \text{salt}$$

$$2HNO_3 + Ba(OH)_2 = Ba(NO_3)_2 + 2H_2O \qquad (16)$$
$$\text{acid} \qquad\quad \text{base} \qquad\quad\quad \text{salt}$$

A more general definition of acid and base is useful in many more complicated reactions and is the Bronsted-Lowry definition:

- An acid is a substance that donates a proton.
- A base is a substance that accepts a proton.

Thus, when HCl dissolves in water:

$$HCl + H_2O = H_3O^+ + Cl^- \qquad (17)$$
$$\text{acid} \quad \text{base}$$

the HCl is an acid and the water is a base, because it accepts the proton. The species formed when a proton is removed from an acid is called the *conjugate base* of the acid; in Equation 17, it is the chloride ion Cl^-, and the species formed when the proton is added to the base is called the *conjugate acid*, which is H_3O^+ for H_2O and HCl for Cl^- in Equation 17.

The equilibrium constant for reaction 17 is

$$K = \frac{[H_3O^+][Cl^-]}{[HCl][H_2O]} \qquad (18)$$

in which the brackets designate concentration; since the water concentration is large and essentially constant, it can be factored out. The remaining factor is called the dissociation or ionization constant K_a of the acid:

$$K_a = \frac{[H_3O^+][Cl^-]}{[HCl]} \tag{19}$$

For some acids such as HCl, HNO_3, and H_2SO_4 the dissociation constant is large, so that the ionization of the acid is almost complete; these are called strong acids. If the dissociation is only partial, the acid is called a weak acid. Examples of acid ionization constants in water are given in Table 7-1. The value of $pK_a = -\log_{10}K_a$ is also given in the table.

In a similar way for a base such as ammonia:

$$NH_3 + H_2O = NH_4^+ + OH^- \tag{20}$$

with an ionization constant

$$K_b = \frac{[NH_4^+][OH^-]}{[NH_3]} \tag{21}$$

For NH_3, $K_b = 1.8(10)^{-5}$ moles/liter, so $pK_b = 4.7$.

7.6 ➤ IONIZATION OF WATER AND pH

Water itself ionizes slightly to form hydrogen and hydroxyl ions:

$$H_2O = H^+ + OH^- \tag{22}$$

The ionization constant for this reaction is

$$K_W = [H^+][OH^-] \tag{23}$$

at 25°C the value of K_w is $1.0(10)^{-14}$ (moles/liter)^2. The concentration units of substances in aqueous solution are often the older units moles/liter. From this value of K_w and Equation 22 the concentration of H and OH$^-$ ions in pure water at 25°C is $1.0(10)^{-7}$ moles/liter.

Because the concentrations of H^+ and OH^- can vary over a wide range, it is convenient to describe acidity in terms of the logarithm of the hydrogen ion concentration:

$$pH = -\log_{10}[H^+] \tag{24}$$

or

$$[H^+] = 10^{-pH} \tag{25}$$

Thus, the pH ("power of hydrogen") is a measure of acidity (or basicity) of a solution. Pure "neutral" water has a pH of 7.0 at 25°C; if the pH >7, the solution is basic, if pH <7, it is acidic. A solution of one mole/liter of hydrogen ions, say from a strong acid, has a pH of zero. The hydroxyl ion concentration is directly related to pH; in water at 25°C

$$[OH] = 10^{ph-14} \tag{26}$$

Thus, if the pH is 9 (basic), the hydroxyl ion concentration is 10^{-5} moles/liter. The pH of some common substances is given in Table 7-2.

7.7 ➤ BUFFERS

Consider a 0.1 m aqueous solution of acetic acid, which is a weak acid. The ionization of acetic acid can be written as

$$CH_3COOH = H^+ + CH_3COO^- \tag{27}$$

with an ionization constant

$$K_a = \frac{[H^+][CH_3COO^-]}{[CH_3COOH]} \tag{28}$$

The value of K_a is about $1.8(10)^{-5}$ moles/liter, or $pK_a = 4.79$.

At $[CH_3COOH] = .1$, $[H^+] = \sqrt{1.8(10)^{-6}} = 1.3(10)^{-3}$

or pH = 2.89. If a few milliliters of one molar HCl solution is added to the acetic acid solution, the pH of the solution will drop sharply (become more acidic) because the hydrogen ions from the HCl stay in solution. A small number of them will react with acetate ions (Equation 27 going to the left), but because acetic acid is a weak base, the concentration of acetate ions is small, in fact equal to $1.3(10)^{-3}$ molar, so the extent of this reaction is small.

Now consider what happens if the concentration of acetate ions in the solution is increased by adding a salt of acetic acid (an acetate, such as sodium acetate) to the solution. Let the concentration of acetate ions be .1M. Now, if some HCl solution is added to this solution, the pH will hardly change at all. The reason is that the added hydrogen ions push reaction 27 to the left, making neutral ionized acetic acid CH_3COOH. The pH of the solution stays about the same (4.74, as calculated below) as long as the acetate ion concentration remains close to .1M. This kind of solution that holds its pH nearly constant even when acid or base is added to it is called a *buffer*. The pH of the solution containing .1M CH_3COO^- and .1M CH_3COOH is 4.74, since $K_a = [H^+][CH_3COO^-]/CH_3COOH = [H^+]$ and $pK_a = 4.74$.

Buffer solutions can be made to cover a wide range of pH by choosing the appropriate weak acid or base. The pH of the buffer solution is given by the Henderson-Hasselbalch equation

$$pH = pK_a + \frac{\log[A^-]}{[HA]} \qquad (30)$$

in which $[A^-]$ and $[HA]$ are the concentrations of ionized and unionized acid, respectively. The pH values of some buffers are given in Table 7-3.

Buffering action also occurs during a titration, in which an acid is neutralized by a base. Let us titrate 25ml of 0.1M acetic acid solution (with no added acetate ion) with 25ml of 0.1M sodium hydroxide solution. The titration curve is given in Figure 7-1. The original solution has a pH of 2.9, as calculated above, just after Eq. 28. As NaOH, a strong base, is added the pH first increases rapidly, but then increases less as acetate ions are added to the solution because of the neutralization of the acetic acid:

$$CH_3COOH + OH^- = CH_3COO^- + H_2O \qquad (31)$$

There is a broad region of pH over which the pH does not change much as NaOH is added, because of the buffering action of the CH_3COOH.

7.8 ➤ MEASUREMENT OF pH

There are two main ways to measure pH. Certain substances have characteristic colors at particular pH values or in ranges of pH values. An example is phenolphthalein, a complex organic molecule. In acid solution, it is colorless, and it is red in basic solution; the exact pH where the dissolved phenolphthalein is half in the colorless and half in the red form is 9.5. There are a range of different indicators that change colors at different pH values, as shown in Table 7-4. These indicators can be incorporated into strips of paper that can be dipped into a solution to give a rough idea of the pH of the solution.

A more accurate way to measure pH is from the electrical potential of a glass electrode. Sodium ions in the glass (*gl*) exchange with hydrogen ions from the solution (*sol*):

$$Na^+(gl) + H_3O^+(sol) = Na^+(sol) + H_3O^+(gl) \qquad \textbf{(32)}$$

making the potential across the glass membrane directly related to the hydrogen ion concentration in solution. This electrode is sensitive over a pH range of about 0 to 10, and can measure pH to ± .01 unit.

The discussions of this chapter show how important chemical equilibria are in many different kinds of chemical problems. Equilibrium constants for a vast number of chemical reactions can be calculated from thermodynamic data, as described in Chapter 9 later in this book.

ENERGY BALANCES—
FIRST LAW OF THERMODYNAMICS

8.1 ➤ INTRODUCTION

Thermodynamics is a physical science concerned with energy changes that accompany transfer of heat and work in physical and chemical processes. Thermodynamics is broadly applicable to all kinds of these processes, such as chemical reactions, performance of engines, electrical processes, and even cosmological events. In this chapter the basics of thermodynamics are introduced and applied to energy balances in these processes, resulting in the first law of thermodynamics. In a subsequent chapter thermodynamics will be extended to more advanced concepts, such as entropy and free energy, that are highly useful in studying chemical and physical processes, especially equilibrium.

8.2 ➤ TEMPERATURE

We all have an intuitive sense of temperature. To determine a quantitative scale of temperature we can use the behavior of ideal gases, as described in the chapter on gases. Let us consider the volume of an ideal gas between 0° and 100°C; that is, the freezing and boiling points of water. The normalized change of volume of the gas is:

$$\frac{V_{100} - V_0}{V_0} = \frac{1}{A} \tag{1}$$

over this temperature range. In this equation V_0 is the volume of gas at 0°C, and V_{100} is the volume of the same mass of gas at 100°C. The constant A is the same for all gases, and is equal to 2.7315. Thus, the volume of any dilute gas can define a temperature scale. Furthermore, extrapolation of the temperature to the point where the volume would be zero gives a temperature of -273.15°C, which is, therefore, the absolute zero of temperature, and can be used to define a new or absolute (Kelvin) scale of temperature:

$$T = -273.15 \times \left(\frac{V}{V_0} - 1 \right) \tag{2}$$

in which V is the volume of gas at temperature T.

8.3 ➤ SOME DEFINITIONS

A thermodynamic *system* is a certain portion of the universe bounded by real or imaginary boundaries; it is, therefore, separated from the rest of the universe, which is called the *surroundings*. For example, the system might be a block of metal that would be subject to heating or cooling, compressing, or an electrical current from the surroundings.

The *state* of a system is determined when we have specified a certain number of independent variables that characterize the system with respect to the features that influence our thermodynamic considerations. Again consider a block of metal. Normally, the temperature and pressure (stress) define the state of the metal. However, in certain considerations, other variables might be important, such as the electrical potential or the magnetic field, or the gravitational field, or the surface area of the sample. Thus, the variables required to define the state of a system depend upon the problem at hand.

An *isothermal* process is one that takes place at constant and uniform temperature. An *adiabatic* process is one that takes place without flow of heat across the boundary of the system.

8.4 ➤ HEAT

Thermal equilibrium is achieved when a system reaches constant and uniform temperature. If two bodies at different temperatures are placed in contact, they eventually reach thermal equilibrium by the flow of heat between them. Thus, heat is the energy that is transferred as a result of temperature differences. The heat Q of a process is considered to be positive when it is added to a system, and negative when it is given off from a system to the surroundings.

8.5 ➤ WORK

There are many different kinds of work. The most familiar is mechanical work that involves the operation of a force through a distance. In three dimensions, mechanical work involves pressure inducing a change in volume. A simple example is a gas in a cylinder being pushed on by a piston. If the gas is at pressure P, and the cylinder has cross-section A, the differential work dW done by moving the piston a differential distance dL is

$$dW = PAdL \qquad (3)$$

or

$$dW = PdV \qquad (4)$$

where dV is the differential volume change, since dV = AdL. In Equation 3, P is not necessarily a constant, because it can be changed by changes in volume.

An important property of the work is that its value depends upon the way the process is carried out. For example, if the process is carried out at constant pressure, then the work

$$W = P(V_2-V_1) \qquad (5)$$

where V_1 is the initial volume and V_2 the final volume. In this process the temperature of the gas must change by heating or cooling from the surroundings to maintain the pressure constant. Another path possibility is isothermal, or constant absolute temperature T. In this case the work is

$$W = RT \ln\left(\frac{V_2}{V_1}\right) \qquad (6)$$

in which R is the gas constant. Equations 5 and 6 can give quite different values of work for the same values of V_2 and V_1. Therefore, the amount of work depends upon the way the process is carried out, not just on the initial and final states of the gas. Thus, work is called a *path* process, not a state function. Although it is not so easy to see, the heat transferred in a process also depends the way the process is done, so heat is also a path-dependent function.

Some examples of other kinds of work are surface work, done when the surface area of a material is changed, electrical work done when a current is passed through a material, magnetic work, and gravitational work.

By convention, the work done on the surroundings by a system is considered positive. This leads to a different convention for heat and work that is unfortunate, but is so deeply entrenched that we must simply go along with it.

8.6 ➤ FIRST LAW OF THERMODYNAMICS

Although both the heat and work of a process depend on the conditions or path of the process, experiments show that their difference does not depend on the path. This result is the first law of thermodynamics: *When a system undergoes a change of state the quantity Q-W depends only on the initial and final states of the system and is independent of the path.* The state function Q-W is given the name internal energy with the symbol U, so for any thermodynamic process the change ΔU is

$$\Delta U = Q - W \qquad (7)$$

This relation shows that Q and W are not completely independent of one another. The function U depends only upon the state of a system, for example, the temperature and pressure and other defining variables, and not on the history of the system. Thus, with a particular set of defining variables, the system will always have the same internal energy U. This result is of great importance, because it allows us to determine this energy without knowing the history of the system. Some examples that show the power of this law will be given below.

Since the definition of the internal energy U depends only on changes in the system, it is not possible to specify an absolute value of the energy from the first law. However, this uncertainty is not important, because the applications of the first law of interest all involve changes of state.

The internal energy U is closely related to energies familiar from particle dynamics. We cannot give more than a short hint of this relation here. In a gas, for example, the kinetic energy of the gas involves the velocity of the gas molecules. As the temperature is increased, the molecules move more quickly, and the kinetic energy increases. In a related way, the internal energy of the gas increases as the temperature increases, as described below.

If a process is carried out at constant volume, the mechanical work is zero, from Equation 4. Thus, the heat Q_v associated with the process is just equal to ΔU. For a constant pressure process, the work equals $P\Delta V$, in which ΔV is the volume change, and the heat Q_p is

$$Q_p = \Delta U + P\Delta V \tag{8}$$

if there is mechanical work only. Most processes of interest take place at constant pressure, so it is convenient to define an additional function called the *enthalpy* H:

$$H = U + PV \tag{9}$$

The enthalpy is also a state function like the internal energy, because all the quantities in its definition (U, P and V) are also state functions. For a constant pressure process

$$Q_p = \Delta H \tag{10}$$

that is, the enthalpy change of a constant pressure process is just equal to the heat.

The units of heat, work, internal energy and enthalpy are all those of energy, or joules. They are called *extensive* properties, because they depend upon the size or extent of the system. Often for convenience the internal energy and enthalpy are given per unit mass or moles of a system; then they become *intensive* properties that do not depend on the size of a system, but only on its defining variables. These intensive units are usually joules/mole.

8.7 ➤ HEAT CAPACITY

The quantity of heat required to raise the temperature of a material a certain amount is an important thermodynamic property that is given the name heat capacity C. Since the heat of a process depends upon the path or way the process is carried out, the heat capacity also depends on the path. A general and rigorous definition of the heat capacity is

$$C = \frac{dQ}{dT} \tag{11}$$

in which dQ is the heat change required to change the temperature of a material by dT. In fact, the heat capacity can vary from minus to plus infinity depending upon the way the process is carried out. However, to be useful the heat capacity is usually defined for a specific condition such as constant pressure. At constant pressure the heat capacity C_p is just the enthalpy change ΔH required to change the temperature of the material by ΔT:

$$C_p = \frac{\Delta H}{\Delta T} = \left(\frac{\partial H}{\partial T}\right)_p \tag{12}$$

The partial differential is a more precise way to define the heat capacity so that it becomes an unique function of the system variables. To compare heat capacities of different materials, a fixed mass or molar amount is taken, so that the units of heat capacity are joules per gram per degree, or more commonly joules per mole per degree.

Some heat capacities of different materials are given in Table 8-1 as joules per mole per degree.

8.8 ➤ HEATS OF REACTION AND FORMATION

Chemical reactions can give off or absorb heat. Burning of a fuel such as methane (CH_4) gives off heat:

$$CH_4 + 2O_2 = CO_2 + 2H_2O \tag{13}$$

whereas the reaction of hydrogen with solid iodine to give gaseous HI absorbs heat:

$$H_2 + I_2 = 2HI \tag{14}$$

Chemical reactions such as 13 and 14 are usually studied at constant temperature and pressure, although in practical applications, the temperature usually changes. Furthermore, to obtain reproducible results, the process (reaction) is considered to take place *reversibly*, which means that it takes place close to equilibrium. To visualize such an experiment, think of a system containing the reactants and products of the reaction close to equilibrium at constant total pressure of reaction. Then as the reaction proceeds, heat must be either added or removed from the system to keep it at constant temperature. Since the pressure is constant, this heat change is just equal to the enthalpy change of the reaction. For example, the enthalpy change for reaction 13 is -890,400 joules per mole of methane burned at 25°C, and one atm pressure. When heat is given off in the reaction, it is called *exothermic*, and the enthalpy change is negative, because heat must be removed from the system to keep it at constant temperature. The enthalpy change for reaction 14 is +52,200 joules per mole of HI formed, and is called *endothermic* because heat must be added from the surroundings to keep the temperature of the system constant. These enthalpy changes are often called the *heats of reaction* for the pertinent chemical reactions.

The *heat of formation* of a substance such as a chemical compound is defined as the heat of reaction of the compound formed from its elements at a particular temperature and pressure. Thus, the heats of formation of gaseous methane and hydrogen iodide at 25°C and one atm. are -890.4 kj/mole and +52.2 kj/mole. With this definition, the heat of formation of an element in its standard state is zero.

Most compounds cannot actually be formed by direct combination of their elements, but because the enthalpy is a state function (first law of thermodynamics), the heat of formation of a compound is still a meaningful concept. The heat of formation can usually be determined indirectly from other experimental reactions. For example, consider the sugar sucrose, $C_{12}H_{22}O_{11}$. Although sucrose cannot be made directly from carbon, hydrogen and oxygen, its heat of formation can be calculated from the heat of the reaction (burning) of sucrose to carbon dioxide and water, and the heats of formation of these two compounds. The following sequence of reactions shows how this calculation is made:

<u>Heat of reaction, kj/mole</u>

$$C_{12}H_{22}O_{11} + 12O_2 = 11H_2O + 12CO_2 \qquad \text{-5637.4} \qquad \textbf{(15)}$$

$$12C + 12O_2 = 12CO_2 \qquad \text{-4729.3} \qquad \textbf{(16)}$$

$$11H_2 + \frac{11}{2}O_2 = 11H_2O \qquad \text{-3145.9} \qquad \textbf{(17)}$$

If the first reaction is reversed and the sum of the three taken, the result is:

$$12C + 11H_2 + \frac{11}{2}O_2 = C_{12}H_{22}O_{11} \qquad \textbf{(18)}$$

with a heat of reaction (formation of sucrose) of -2232.8 kj/mole.

By experiments and calculation similar to the ones just described, the heats of formation of a large number of compounds are available. A sampling is in Table 8-2. In references given at the end of the chapter, there are lists of large numbers of heats of formation. From these values the heats of reaction of reactions whose heats have not been measured experimentally can be reliably calculated. This procedure illustrates the great value of the first law of thermodynamics. Because enthalpy is a state function that does not depend upon the history of a process, the final state can be calculated from other experiments. Thus, heats of reactions of any reactions whose heats of formation are known can be calculated; one does not need to measure the heat to get a reliable value.

8.9 ➤ HEATS OF PHASE CHANGES

Heats of phase changes can be treated in a similar way to the heats of reactions. The enthalpy changes for vaporization, melting, sublimation or a structural change of a solid are the heats of these processes at a constant temperature and pressure. For example, the heat of vaporization of water at 100°C and one atm. pressure is 40.7 kj/mole, and is positive to vaporize the water. If water condenses at 100°C, the heat is the same except that it is negative, because it is given off by the transformation.

8.10 ➤ BOND ENTHALPIES

Consider a process in which a chemical compound is completely dissociated into its separate gaseous atoms. The enthalpy change for this process at constant pressure is the sum of the heats of dissociation of all the chemical bonds in the compound. For example, consider the disassociation of gaseous ethane into carbon and hydrogen atoms:

$$C_2H_6 = 2C + 6H \tag{19}$$

The enthalpy change for this reaction is called the *bond enthalpy* of ethane. Often bond enthalpies are called bond energies. The difference between these two quantities is the change in PV (Equation 9) and is usually negligible compared to other uncertainties.

Heats of formation of gaseous atoms have been measured, usually by spectroscopic methods, so the bond enthalpies of compounds can be found from these values and the heat of formation of the compound. The total bond enthalpy of a compound is made up of the total of the individual bond enthalpies in the compound. For ethane the heat of reaction 19 or its bond enthalpy is 2840 kj/mole, composed of six C-H bonds of enthalpy 416 kj/mole each and one C-C bond of enthalpy 343 kj/mole. The structure of ethane is

```
        H  H
        |  |
    H—C—C—H
        |  |
        H  H
```

Studies of the bond enthalpies in many different compounds shows that these enthalpies are about the same in compounds with widely different structures. Thus, useful average values of bond enthalpies can be calculated. Table 8-3 lists average bond enthalpies of many different types of bonds. Such a table is of great value, because it can be used to calculate enthalpies of formation of complicated compounds that have not been measured.

The availability of tables of thermodynamic properties such as heat of formation and bond enthalpies, which can be used to calculate unknown quantities, demonstrate one aspect of the great power and usefulness of the first law of thermodynamics.

Table 8-1 and 8-2
Heat Capacities and Heats of Formation
of Some Important Materials at 25°C

Name	Symbol	Heat Capacity j/mole°K	Heat of Formation kj/mole
GASES Carbon Dioxide	CO_2	37.1	-394
Chlorine	Cl_2	33.9	0
Methane	CH_4	35.7	-74.8
Oxygen	O_2	29.4	0
LIQUIDS Ethyl alcohol	C_2H_5OH	111.5	-277.6
Methyl alcohol	CH_3OH	81.6	-238.6
Water	H_2O	75.3	-285.8
SOLIDS Aluminum	Al	24.3	0
Copper	Cu	24.5	0
Iron	Fe	25.2	0
Diamond	C	6.06	1.90
Graphite	C	8.64	0
Silicon	Si	19.87	0
Salt	NaCl	49.7	-411
Alumina	Al_2O_3	79.0	-1670
Ferric Oxide	Fe_2O_3	104.6	-822
Quartz	SiO_2	44.4	-859

Table 8-3
Bond Enthalpies at 0°K in kj/mole*

	H	C	N	O	S	F	Cl	Br	I
H	436	414	389	464	339	565	431	368	297
C		347	293	351	259	485	331	276	218
N			159	222		272	201	293	
O				138		184	205	201	201
S					226	285	255	213	
F						153	255	255	277
Cl							243	218	209
Br								193	180
I									151

Double Bonds		Triple Bonds	
C=C	612	C≡C	820
C=N	615	C≡N	890
C=O	715	C≡O	1075
C=S	477	N≡N	941
N=N	418		
N=O	607		
O=O	498		
S=O	498		

from W.Y. Masterton and C.N. Hurley: "Chemistry," Saunders, N.Y., p. 214.

9

SECOND LAW OF THERMODYNAMICS

9.1 ➤ INTRODUCTION

In Chapter 8 on energy balances and the first law of thermodynamics, it was shown how heats of reactions and phase transformations can be accurately calculated from tables of enthalpy values or estimated from bond enthalpies. In this chapter the principles of thermodynamics will be extended to determination of equilibrium of chemical reactions and phase transformations. This application is of major importance, because it provides predictions of the extent of chemical reactions to be expected in unknown systems.

There are many other applications of thermodynamics that cannot be discussed in this brief treatment. One of great importance in design of engines, refrigerators, and many other devices is a measure of their efficiency. Historically the science of thermodynamics was first developed to calculate the maximum amount of useful work that can be extracted from a heat engine. Thus the second law of thermodynamics is often stated in terms of the efficiencies of heat engines. However, in chemistry and materials science the great usefulness of the second law is to determine criteria for equilibria. Thus we have chosen to introduce the second law in terms of the properties of the entropy function. The two approaches are equally valid, but we will not prove this conclusion, because it requires many additional ideas and mathematical tools to those we are using here.

9.2 ➤ SECOND LAW OF THERMODYNAMICS

Consider a pure single component material in a particular phase. Then a second law equation for this system is:

$$\Delta U = T\Delta S - W_R \qquad (1)$$

in which ΔU is a change of internal energy U for the phase, T is the temperature, ΔS is the change in S, a new function called <u>entropy</u>, and W_R is the reversible work required to bring about these changes. The second law states that the entropy function S is a state function; that is, that its value depends only on the beginning and ending state of the phase, and not on the path by which the change takes place. This statement of the second law is somewhat similar to the statement of the first law in Chapter 8, in which the internal energy is stated to be a state function. Both these laws of thermodynamics are stated as axioms, so they do not derive from other considerations. Their validity has been established by a vast number of experiments; there is no reliable experimental evidence contrary to these laws.

A phase is defined as a region of material with uniform structure, composition, and properties. In general a phase can contain many components; for example, an alloy of tin dissolved in lead and a solution of sugar in water are single phases. For the purposes of this chapter, we restrict the statement of Eq. 1 to a pure phase; that is, one that contains only a single chemical component (pure gold, pure water, pure salt). Eq. 1 can be extended to multicomponent phases, but this extension requires concepts we will not introduce in this book.

The work term in Eq. 1 can include all kinds of work, but we will consider only mechanical or pressure volume work. The <u>reversible</u> work is the work from a particular idealized path in which changes in state functions (pressure, temperature) can be arbitrarily small and close to equilibrium. The reversible work is a state function, because all the other quantities in Eq. 1 are state functions.

9.3 ➤ ENTROPY

The first law equation (Eq. 7, Chapter 8) for the phase of Eq. 1 is

$$\Delta U = Q_R - W_R \qquad (2)$$

Since both E and W_R are state functions, the <u>reversible</u> heat Q_R from Eq. 2 must also be a state function. Thus both the reversible heat and work for a particular change of state will have only one value. From a comparison of Eqs. 1 and 2

$$\Delta S = \frac{Q_R}{T} \qquad (3)$$

so that the change of entropy for a process can be calculated from the reversible heat of the process. In the next paragraph Eq. 3 is used to calculate entropy changes for two different processes.

A phase transformation such as melting or freezing takes place at the equilibrium temperature, so it is reversible. The heat of fusion ΔH_f in joules/mole is absorbed on melting or given off during freezing, so the entropy of fusion is $\Delta H_f/T_m$, where T_m is the melting temperature. The units of entropy are joules per mole per degree kelvin.

In Chapter 8 the work to expand n moles of an ideal gas at constant temperature T from pressure p_1 to p_2 was found to be $nRT\ln(p_1/p_2)$, where R is the gas constant. The energy U for an ideal gas depends only on temperature, so ΔU for this process is zero. Therefore, from Eq. 2 for this change, $Q_R = W_R$, and the entropy change is equal to $Q_R/T = W_R/T = nR\ln(p_1/p_2)$.

Eq. 1 and the statement that the entropy S is a state function is the first part of the second law of thermodynamics. In addition the condition for an irreversible reaction must be stated. It is: <u>If an isolated system is not at equilibrium, then the entropy of the system is increasing.</u> An isolated system is one for which there is no heat, material, or work exchanged with the surroundings, so that from the first law the energy of the system is constant. Its volume and overall composition are also constant. In symbols

$$\Delta S > 0 \qquad\qquad (4)$$

for constant energy U, volume V, and content, for an irreversible reaction. If the system is at equilibrium,

$$\Delta S = 0 \qquad\qquad (5)$$

for an isolated system.

9.4 ➤ ENTROPY CHANGES FOR CHEMICAL REACTIONS

The entropy change for a chemical reaction can be calculated in a way similar to the calculation of heats of reaction described in Chapter 8. A table of standard entropies is useful for this calculation (see table).

Only differences of energy or enthalpy are used in thermodynamic calculations, because absolute values of these functions cannot be easily calculated. An absolute value of entropy, however, can be calculated as described in appendix. These calculations lead to absolute entropy values for both compounds and elements, and these entropies can be calculated as a function of temperature and pressure. The standard entropy is defined at a particular temperature and pressure, usually 25°C and 1 atm. A tabulation of standard entropies for these conditions is in Table 9-1. Elements have non-zero standard entropies, and the standard entropies of pure substances (elements and compounds) are always positive. Their units are joules/moles - degree.

These standard entropy values can be used to calculate standard entropy changes of chemical reactions. For example, consider the oxidation of iron to ferric oxide:

$$2Fe + 3O_2 = 2Fe_2O_3 \tag{6}$$

The standard entropies of the constituents in this reaction at 25° and 1 atm. are Fe, 27.3; O_2, 205.0; and Fe_2O_3, 87.4.

The standard entropies must be multiplied by the reaction coefficient for each compound. The entropy of reaction is found by subtracting the sum of entropies of reactants from the sum of entropies of products:

$$\Delta S^0 = 174.8 - 54.6 - 410 = -289.8 \ j/mole\text{-}° \tag{7}$$

The entropy change for this reaction is highly negative because of the large entropy of gaseous oxygen. In general, gases have larger entropies than liquids or solids.

9.5 ➤ FREE ENERGY

To study processes such as chemical reactions and phase transformations, experiments are most conveniently done at constant temperature and pressure. Thus, the conditions of constant energy and volume for Eq. 4 are not convenient, and it is necessary to define another function that gives a criterion for equilibrium at constant temperature and pressure. This function is the Gibbs free energy, named for the American physicist Josiah Gibbs, and is defined as:

$$G = H - TS \tag{8}$$

in which H is the enthalpy, S is the entropy, and T is the temperature. Since all these functions are state functions, G is a state function.

At constant temperature and pressure and with mechanical (pressure-volume) work only, it can be shown (see appendix) that the conditions of Eqs. 4 and 5 lead to the condition that for equilibrium the Gibbs free energy change is zero:

$$\Delta G = 0 \quad (equilibrium) \tag{9}$$

and for a process not at equilibrium the Gibbs free energy decreases with time

$$\Delta G < 0 \quad (non\text{-}equilibrium) \tag{10}$$

until equilibrium is reached. Thus, at constant temperature and pressure with only pV work the Gibbs free energy gives a <u>criterion for equilibrium</u>: The Gibbs free energy tends to decrease until it reaches a minimum, at which point the system is in equilibrium.

9.6 ➤ CALCULATION OF FREE ENERGIES OF REACTION

In Chapter 7 the concept of equilibrium of a chemical reaction was discussed. A general equilibrium constant K for reactions with reactants and products that are ideal gases was defined (Eqs. 7-9). The free energy change of a chemical reaction can be used to calculate the equilibrium constant of Eqs. 7-9 and of any other chemical reaction, whether it is homogeneous or heterogenous with ideal or non-ideal gases, and for any phases involved in the reaction. Thus in this section we discuss how to calculate the free energy change of a reaction, and then apply this change to the calculation of the equilibrium constant in the following section.

The method used to calculate free energy changes of chemical reactions is very similar to that used for calculating heats of reaction in Chapter 8. From Eq. 6 the change ΔG in Gibbs free energy at a constant temperature is given by:

$$\Delta G = \Delta H - T\Delta S \tag{11}$$

This equation is sometimes called the Gibbs-Helmholtz equation.

In Chapter 8 we described how heats of reaction are calculated from tables of heats of reaction. Similarly entropies of reactions can be calculated from standard entropies, as described in Section 9.4 above. Thus, Equation 11 can be used to calculate standard free energies of reaction from these two quantities. In practice it is easier to start with free energies of formation, just as done with heats of formation, and to compile a table of free energies of formation of elements and compounds.

The Gibbs free energy of formation of a substance is the standard free energy of reaction to form the substance from its elements in their stable states, at one atm. pressure. For example, consider methane:

$$C + 2H_2 = CH_4 \tag{12}$$

Let us write Eq. 11 with small superzeros to denote standard conditions:

$$\Delta G^0 = \Delta H^0 - T\Delta S^0 \tag{13}$$

Then, from the tables ΔH^0, the heat of formation of methane, is -79.8 kjoules/mole, and the standard entropy change for reaction 12 is 186.2 - 2 x 130.6 - 5.7 = 80.7 joules/mole deg., =.0807 kj/mole deg. If these two quantities are inserted into Eq. 13, ΔG^0 = -50.7 kj/mole. Thus, free energies of formation for substances can be calculated from tables of heats of formation and entropies. Values of Gibbs free energies of formation are given in Table 9-1. From them values of free energies of reaction can be calculated.

Sample Problem:
Calculate the free energy change for the reaction of lead oxide with sulfur dioxide at 25°C:

$$2PbO + 2SO_2 = 2PbS + 3O_2 \tag{14}$$

In this reaction the lead oxide and lead sulfide are solids and the sulfur dioxide and oxygen are gases. From the table of free energies of formation at 25°C and atm:

$$\Delta G^0 = 2 \times \Delta G^0(PbS) - 2 \times \Delta G^0(SO_2) - 2 \times \Delta G^0(PbO) \tag{15}$$

$$\Delta G^0 = -197.5 + 600.4 + 377.8 = +780 \; kj/mole \tag{16}$$

Sample Problem:
Calculate the free energy change for the reaction of ferrous oxide with carbon at 25°C and one atm.

The reaction of iron oxide with carbon to form iron and carbon dioxide is:

$$2Fe_2O_3 + 3C = 4Fe + 3CO_2 \tag{17}$$

$$\Delta G^0 = 3 \times (-394.4) - 2 \times (-742.2) = +301.2 \; kj/mole \tag{18}$$

The exact calculation of free energies of reactions at temperatures other than 25°C requires values of free energies of formation at different temperatures. As an approximation that is valid for our purposes, however, we can assume that ΔH^0 and ΔS^0 are not functions of temperature. Then the only change with temperature is the temperature factor in Eq. 10.

Sample Problem:
Calculate the free energy of reaction for reactions 14 and 17 above at 1000°C.

For this calculation we need values of ΔH^0 and ΔS^0 for the reactions at 25°C (298°K). These are: for reaction 14,

$$\Delta H^0 = -830.8 \; kj/mole \qquad\qquad (19)$$

$$\Delta S^0 = 168 \; j/mole \qquad\qquad (20)$$

for reaction 17:

$$\Delta H^0 = +467.9 \; kj/mole \qquad\qquad (21)$$

for reaction 14 at 1237°K (1000°):

$$\Delta G^0 = 830.8 - 1273(-168) = 617 \; kj/mole \qquad\qquad (22)$$

$$\Delta S^0 = 559 \; j/mole \qquad\qquad (23)$$

for reaction 17 at 1237°K (1000°):

$$\Delta G^0 = +467.9 - 1273(.559) = -244 \; kj/mole \qquad\qquad (24)$$

As mentioned above, these results are approximate.

Thus, the free energy change for the reduction of iron oxide with carbon changes from positive at room temperature to negative at 1000°C, whereas the reaction of lead oxide with sulfur dioxide is positive at both of these temperatures. The implications of these results have great practical importance in the utilization of these reactions, and are explored in the next section.

9.7 ➤ SPONTANEITY OF CHEMICAL REACTIONS AND THE CALCULATION OF EQUILIBRIUM CONSTANTS

In Section 9.5 above we showed that the free energy change for a spontaneous process is negative; if the free energy change is positive, the process does not take place. Thus, the free energy changes of chemical reactions tell whether or not these reactions will take place. In many cases rates of reactions may be slow and limit the actual amount of reaction that takes place. Nevertheless the free energy change of a reaction tells whether it is possible or not. If the change is positive, it will not take place. Rates of reactions increase sharply with temperature, so reactions that are slow at room temperature can become faster at higher temperatures.

Let us examine the reactions for which free energies were calculated in the last section. Eq. 12 for the formation of methane from carbon and hydrogen has a negative free energy change, so it can take place at room temperature. Almost all the free energies of formation in Table 9-1 are negative, showing that the reactions of formation of compounds from their elements are almost always likely. An exception is the formation of acetylene C_2H_2; this result demonstrates the extreme instability of this compound, which can spontaneously decompose with explosive violence under certain conditions. The gaseous nitrogen oxides NO, NO_2 and N_2O_4 are also unstable at room temperature, although they become stable at higher temperatures. Their formation in engines lead to pollution of the atmosphere, and their reactions are strong factors in smog formation. In the body nitrous oxide, NO, is the subject of intense research because of its role in certain physiological reactions, again partly because of its instability.

The free energy of reaction 6 is strongly negative; the formation of iron oxide at room temperature (rust formation) is catalyzed (made more rapid) by water.

The free energy for reaction 17 is positive from room temperature to 1000°C. This means that lead oxide will not react with sulfur dioxide to form lead sulfide in this temperature range. If the reaction is reversed, we find the opposite is true: lead sulfide can react with oxygen to form lead oxide over this temperature range.

The change in sign for reaction 15 shows that carbon dioxide can react with iron to form iron oxide at room temperature; however, the rate of this reaction is very slow at this temperature. At 1000°C, Fe_2O_3 is reduced by carbon to form elemental iron; this is one way of forming iron metal from certain ores, as discussed more completely in Chapter 10.

The free energy of a reaction gives not only a qualitative measure of the extent of a reaction, but also a quantitative one. This quantitative measure comes from the calculation of the equilibrium constant (Chapter 7) from free energies of reaction. The relation between the standard free energy change ΔG^0 of a reaction and its equilibrium constant K from Eq. 9, Chapter 7, is:

$$\Delta G^0 = RT\ln K \qquad (25)$$

in which R is the gas constant (8.31 joules/moles-deg), and T the absolute temperature. We will not derive this equation because it is beyond the scope of this book, but we will show how it can be used to find exactly how much a mixture of substances will react.

Let us calculate K for reaction 17 of iron oxide with carbon. Since pure solids are considered at unit concentration for calculating K,

$$K = P_D^3 \qquad (26)$$

in which P_D is the pressure of carbon dioxide at equilibrium for reaction 17. The value of K at 1000°C is:

$$K = \exp\left(\frac{-\Delta G^0}{RT}\right) \qquad (27)$$

$$K = \exp\left(\frac{+244,000}{8.31}x1273\right) \qquad (28)$$

$$K = \frac{\exp(+23.06)}{[1.03(10)^{10}]^{\frac{1}{3}}} = 1.03(10)^{10} \qquad (29)$$

Thus, $P_D = 2.18(10)^3$ atm. in equilibrium with Fe_2O_3, iron, and carbon at 1000°C. If the pressure of carbon dioxide in the reaction is above $2.18(10)^3$ atm. the reaction goes to the left, so that carbon dioxide reacts with iron to form iron oxide and carbon. If the carbon dioxide pressure is less than $2.18(10)^3$ atm., the iron oxide is reduced to iron.

In Chapter 7 the equilibrium constant for the reaction of carbon dioxide with hydrogen to form carbon monoxide and water (reaction 7-1) was stated to be 0.39. Let us calculate this value from the table of free energies of formation, Eq. 13 and Eq. 25. The standard free energy ΔG^0 for this reaction at 25°C is:

$$\Delta G^0 = -228.6 - 137.2 - (-394.4) = +28.6 \; kj/mole \qquad (30)$$
$$\text{water} \qquad CO \qquad CO_2$$

Recall that the free energy of formation of elements such as hydrogen is zero. To estimate ΔG^0 at 600°C, use

$$\Delta H^0 = +41.2 \; kj/mole \; at \; 25°C$$

and

$$\Delta S^0 = .0421 \; kj/mole$$

Then, at 600°C:

$$\Delta G^0 = 41.2 - 873(.0421) = +4.4 \; kj/mole$$

Then, from Eq. 25:

$$K = \exp\left(\frac{-4,400}{8.31} x873\right)$$

$$K = \exp(-.607) = 0.54$$

The difference between this result and the exact value of 0.39 comes about because of the neglect of the temperature dependencies of ΔH^0 and ΔS^0 in this method.

9.8 ➤ ENTROPY AND DISORDER

The discussions of thermodynamics in this chapter and in Chapter 8 have so far been concerned with macroscopic phenomena only; there has been no need to introduce atoms or molecules in the discussions. Much of thermodynamics was developed before atoms were generally accepted as the building blocks of matter, and it lost nothing as a result.

Nevertheless, it is tempting to relate the ideas of macroscopic thermodynamics that have been so successful in predicting the possibility and extent of chemical reactions to atomic and molecular phenomena. The science of statistical mechanics has been developed to connect these phenomena to thermodynamics, and in certain carefully defined and restricted conditions leads to additional insight into thermodynamic functions.

In addition to statistical mechanics there have arisen some interpretations of thermodynamic functions in molecular terms that have been extended far beyond their realm of validity. Some of these ideas are considered in this section and the next one. The conclusion is that these ideas are correct only under certain special conditions, and do not have the general validity of thermodynamic methods already discussed.

Many scientists and engineers have been taught that the increase of entropy of a thermodynamic system means that the atoms and molecules of the system are increasing in "disorder" or "randomness" or "mixed-upness." These terms are not exact, but can be interpreted for certain systems. For example, consider two gases in separate chambers. When they are mixed in a single chamber they become more mixed up. Consider a crystal of salt and a glass of water. When the salt is dissolved in the water, these two substances are more mixed up.

The relation between entropy and mixing usually assumed, that is, that an increase in entropy leads to an increase in mixing, is valid only in certain carefully defined situations. It is not a generally valid principle, because there are many exceptions to it. In an isolated system (one with no heat, work, or material exchanges with the surroundings) of non-interacting particles, such as an ideal gas at low temperature and pressure, an increase in mixing does lead to an increase in entropy. In many other systems, however, there is no simple molecular interpretation of changes in entropy.

This subject has been carefully discussed by M. L. McGlashan in his book *Classical Thermodynamics* (Academic Press, 1979), where he emphasizes that "...the entropy change is capable of no simple geometrical interpretation <u>even for changes in isolated systems</u>." (p. 13)

An example of a process in which a system clearly becomes less mixed up yet accompanies an increase in entropy comes from the solution of salts in water. Most substances have increased solubility in water as the temperature increases, for example, sugar or salt. Consider a system consisting of a saturated solution of sugar or salt in water that also contains some solid salt or sugar. As the system is reversibly heated a small amount, more salt or sugar dissolves. The entropy of the system increases, and it is more mixed up. Now consider a saturated solution of sodium sulfate (Na_2SO_4) containing some crystals of sodium sulfate. The solubility of this salt decreases as the temperature is increased, so if the system is heated reversibly some crystals of sodium sulfate precipitate from the solution. The entropy increases, but the system becomes <u>less</u> mixed up. There are many other examples of a decrease of disorder accompanying an increase of entropy.

The conclusion of this discussion, as mentioned above, is that one must be very careful in relating mixing of atoms and molecules (randomness or disorder) to entropy changes. In systems with non-interacting particles an increase of mixing does lead to an increase in entropy, but in most ordinary and practical systems there is no simple interpretation of entropy changes.

9.9 ➤ PHILOSOPHY

The great power and generality of thermodynamics has led many scientists and non-scientists to philosophical speculation about the "meaning" of thermodynamics. One such speculation derives from the supposed relation of entropy to disorder described in the last section. This relation has been extrapolated to realms in which there is no justification for it, such as the "universe", and has been given an almost mystical quality. As McGlashan emphasizes, "Thermodynamics is an experimental science, and not a branch of metaphysics" (loc. cit., p. 111). The application of thermodynamics to the "universe" has been especially common. But for thermodynamics to have any meaning, there must be a system and a surroundings. What are the surroundings of the universe? It is wisest to restrict thermodynamics to systems that can be measured experimentally, and not to try to endow it with any more meaning than the prediction of highly important experimental quantities.

Table 9-1
Standard Entropies and Gibbs Free Energies of Formation
of Some Important Materials at 25°C and One Atm. Pressure

Name	Symbol	Standard Entropy j/mole°K	Free Energy of Formation kj/mole
GASES			
Carbon Dioxide	CO_2	214	-394
Chlorine	Cl_2	223	0
Methane	CH_4	186	-50.7
Oxygen	O_2	205	0
LIQUIDS			
Ethyl alcohol	C_2H_5OH	161	-174.9
Methyl alcohol	CH_3OH	127	-166.3
Water	H_2O	70	-285.8
SOLIDS			
Aluminum	Al	28.3	0
Copper	Cu	33.2	0
Iron	Fe	27.3	0
Diamond	C	2.44	2.87
Graphite	C	5.69	0
Silicon	Si	18.8	0
Salt	NaCl	72.4	-384
Alumina	Al_2O_3	50.9	-1582
Ferric Oxide	Fe_2O_3	87.4	-742
Quartz	SiO_2	41.8	-805

10

PROCESSING OF MATERIALS

10.1 ➤ INTRODUCTION

Solid materials are vital to many parts of technological society. The progress of civilizations is often marked by the development of these materials (Stone Age, Bronze Age, Iron Age). In this chapter, the main categories of these materials and their uses are listed, and then ways of making useful materials are described. Progress in many areas of technology depends on development of improved materials. Examples are: more efficient engines operating at higher temperature, more dense and faster electronic circuits, improved catalysts for making chemicals, and substitutes and repair of human organs and parts. Engineers know how to design improvements in these areas, but materials to make these improvements possible are not yet available.

The main categories of solid materials are ceramics and glasses, electronic materials, metals and alloys, polymers, and combinations, called composites. Example of each category are given in Table 10-1, and uses are listed in Table 10-2.

In this chapter, there are sections on the making and forming (processing) of metals, ceramics, glasses and electronic materials (semiconductors). The discussion of processing of polymers is delayed until Chapter ___, because knowledge of polymer processing requires understanding of organic reactions, which will be introduced in Chapter ___.

10.2 ➤ CERAMICS

Ceramics are the oldest materials processed by humans. Pottery was made far before there was written history. The raw material for pottery is clay (see Table 10-3), which is a common part of many soils. Forming a damp clay into a desired shape, drying it, and firing it at a high temperature are the processing steps for pottery, and also for most other ceramic materials. The sequence form-dry-fire is the usual one for a wide variety of ceramics.

Some common raw materials for ceramics are listed in Table 10-1. Most of these materials are found reasonably pure in nature, and are often used without further purification. Use of these raw materials is economical, but can lead to variability in quality of the final product. For certain "fine" ceramics, such as alumina, the raw materials must be carefully purified to remove common impurities such as silica (SiO_2) and iron oxide (Fe_2O_3) in bauxite ($Al(OH)_3$).

The raw materials must be prepared as powders of desired particle size and size distribution by grinding and then separation and blending. The slightly damp powder is then formed into the desired shape, usually by pressing in a mold. Extrusion of the powder through a die is also possible; usually the powder contains an organic "binder" to hold it together. Forming by hand is still common for art ware.

The formed ware is then dried, usually at about 100°C, to prevent excessive shrinkage and explosion of the material when it is fired at high temperature. Finally the ware is fired, usually at temperatures from about 800° to 1400°C. Higher temperatures are used for special materials. During firing the individual powder particles are joined together in a process called <u>sintering</u>. In most ceramics there are some low melting components present, such as oxides containing alkali metals (e.g., feldspar), so that is a liquid phase present during firing. This liquid phase coats the solid particles, and helps to speed the transport of materials to join together the solid particles.

In certain pure materials, such as alumina, the melting temperature is much higher than convenient and economical firing temperatures. In this case, the powder particles join together as solids without any liquid phase present. This solid state sintering is much slower than liquid phase sintering, because material transport is much faster in the liquid.

In recent years, some new methods have been developed to make ceramic materials of high value. These methods are much more expensive than the traditional methods of processing ceramics, because they require extensive preparation of special starting materials. However, they make possible forming of entirely new materials, better control of the final product, and lower forming temperatures. An example is sol-gel formation of ceramics. In this method organic compounds containing the ceramic oxides, such as tetrapropyl titanate, $Ti(OC_3H_7)_4$, and tetraethyl silicate, $Si(OC_2H_3)_4$, are mixed as liquids at or near room temperature. This mixing gives a uniform and homogenous starting material, which is then carefully heated to remove water and the organic material, giving a highly homogenous sintered solid at lower firing temperature than normally required.

10.3 ➤ GLASSES

Silicate glasses have been made by humans for ten thousand years at least. They were originally used for decorative purposes and were highly prized. About two thousand years ago, Roman artisans learned how to blow glass, and its uses expanded, at first as containers and later as windows. The most used glasses are soda-lime silicates, and their composition has changed remarkably little over the millennia. Their raw materials are sand (SiO_2), soda ash (Na_2CO_3), and limestone ($CaCO_3$) or dolomite ($CaMg(CO_3)_2$), mixed to give a final composition of about 75% SiO_2, 15% Na_2O, and 10% CaO + MgO. This glass is relatively easy to melt and form, and is strong, optically clear, and chemically durable. Commercial soda-lime glasses contain a large variety of impurities, some coming with the raw materials. About 2% alumina is often added to increase chemical durability, and small amounts of Sb_2O_3, As_2O_3, or sulfate are added to help remove bubbles during melting. Small amounts of metal oxides impart color to glass: Chromium, green; cobalt or copper, blue; manganese, yellow or purple; iron, green or brown. Finely divided metal particles also give colors: gold and copper, red; and silver, yellow.

Other important glass compositions are pyrex borosilicate (81% SiO_2, 13% B_2O_3, 4% Na_2O, 2% Al_2O_3), which is more expensive because it requires higher melting temperature and borax is a relatively expensive raw material. This glass has good thermal shock resistance, so it is useful for cookware, laboratory glassware, and automobile headlamps. Silicate glass containing considerable amounts (8 to 40%) of lead oxide has a lower softening temperature, so it is useful for sealing together glasses and ceramics. Lead glass also has a higher refractive index to light, so it is used for glass "crystal" and jewelry because is sparkles. Pure silica glass is expensive because it must be melted at high temperature, above 1740°C, the melting temperature of crystalline cristobalite (SiO_2). Silica glass is used for lamps and optics because of its wide range of optical absorption and its excellent thermal shock resistance because of a very low coefficient of thermal expansion. It is also used for semiconductor crucibles because it is so pure.

One of the great triumphs of recent technology was the development of fiber optic cables made of silica glass. These cables are rapidly replacing metal communication lines. They can carry much more information per unit weight, and are spy-proof, because their signal cannot be sensed externally. These cables are about 10μm in diameter, and are up to a kilometer in length. The glass must be extremely pure to prevent optical absorption losses over distances of kilometers, and extremely strong to prevent them from breaking. They were developed by a cooperative research program at Bell Telephone Laboratories and Corning Glass Co.

Soda-lime glass is melted commercially in large "tanks," for example with dimensions 20 meters long, 8 meters wide and 5 meters deep or larger. The raw materials are fed in as powders at one end, and the tank is heated by gas or oil jets across the surface of the molten glass. The glass flows through the tank and becomes homogenous, and bubbles are removed. The glass can be formed into different shapes as it leaves the tank.

A layer of molten glass is floated into a bath of molten tin to make flat glass for windows. This remarkable float process was developed about thirty years ago by Alastair Pilkington in England, and is a great improvement (it is much cheaper) over previous processes that required extensive polishing of the glass surfaces. Gobs of glass are molded into shapes for containers. A thin ribbon of glass can be blown into lamp bulbs by gas jets. These processes are all continuous, just like an assembly line, and are consequently economical. Glasses can also be melted in batches in crucibles for special compositions and smaller amounts, and for hand glass blowing.

Fiber glass is rapidly drawn through a die at the bottom of a special furnace. Usually the glass is fed into the furnace as pre-melted marbles, to give homogeneity.

10.4 ➤ METALS

The stages of civilization are often defined in terms of the predominant metals used in them. We might define our stage as an aluminum age, because of the recent rapid increase in the use of this metal. Nevertheless iron and steel are still by far the most used metals by tonnage. Metals are absolutely vital to our technology, as emphasized by the sampling of uses in Table 10-2.

Historically the metals first used by humans were those found unreacted or "native" in nature. These metals include gold, silver, copper, mercury and platinum.

Most metals are naturally combined in compounds such as chlorides, oxides, and sulfides; these compounds are mixed with silicate rocks as ores. The first step in making these metals is to concentrate the metallic compound by separating it from the silicate. One method for this separation is called floatation, in which the ore is ground up and mixed with water containing a detergent. Air is bubbled through this mixture; the detergent preferentially wets the particles of the metallic compounds, which float to the top of the mixture and are skimmed off and dried. The rock particles sink to the bottom.

The separated metallic compounds are then treated by various reduction reactions to reduce them to metal, as listed in Table 4. Some of these methods are described in more detail below; electrolytic processes will be considered in the chapter on electrochemistry.

Some sulfide compounds such as Cuprous Sulfide, Cu_2S, can react with oxygen (air) directly to form the metal:

$$Cu_2S + O_2 = 2Cu + SO_2 \qquad (1)$$

This reaction is carried out by heating the Cu_2S in air at about 1100° to 1200°, or blowing air or oxygen through the powdered compound. Since copper melts at 1083°C, the metallic copper that forms is molten. The resulting "blister copper" that forms on cooling has voids from air bubbles and contains some impurities. A more reactive sulfide such as ZnS must first be reacted to the oxide:

$$2ZnS + 3O_2 = 2ZnO + 2SO_2 \qquad (2)$$

and then reduced with carbon:

$$2ZnO + C = 2Zn + CO_2 \qquad (3)$$

One might suppose that lead sulfide could also be reacted with air to form lead oxide:

$$2PbS + 3O_2 = 2PbO + 2SO_2 \qquad (4)$$

This reaction is considered in Chapter 9 (Eq. 9.14) on thermodynamics. It is shown that at all temperatures from room temperature to 1000°C, the PbS is the stable lead compound if the gases are at one atmosphere. Thus, thermodynamics shows that this reaction takes place only if the pressure of SO_2 is reduced by removing it during the reaction.

Many metals are found naturally as oxides. Iron is found mainly as hematite (Fe_2O_3) and magnetite (Fe_2O_3-FeO). These compounds are reduced commercially by carbon in a blast furnace, see Fig. 21.3, p. 812, Chang. A mixture of iron oxide, coke (carbon)and limestone ($CaCO_3$) is fed into the top of the furnace. Air or oxygen is fed into the furnace from the bottom. The following reactions occur:

1. The coke burns to carbon dioxide in the bottom of the furnace, which rises and is reduced to carbon monoxide.

$$C + O_2 = CO_2 \qquad\qquad (5)$$

$$CO_2 + C = 2CO \qquad\qquad (6)$$

The net reaction is

$$2C + O_2 = 2CO \qquad\qquad (7)$$

which gives off much heat. This heat helps to keep the bottom of the furnace at a high operating temperature of about 1600°C, above the melting point of iron of 1540°C.

2. The iron oxides are reduced to metallic iron by carbon monoxide:

$$Fe_2O_3 + 3CO = 2Fe + 3CO_2 \qquad\qquad (8)$$

The molten iron is collected at the bottom of the furnace and periodically drawn off.

3. The limestone decomposes to calcium oxide at the top of the furnace:

$$CaCO_3 = CaO + CO_2 \qquad\qquad (9)$$

The calcium oxide is highly reactive, and reacts with silicate impurities in the iron oxides at the top of the furnace:

$$CaO + SiO_2 = CaSiO_3 \qquad\qquad (10)$$

This calcium silicate falls through the furnace and melts to form liquid or slag, which is less dense than molten iron and so floats on its surface.

The solidified iron from the blast furnace is called pig iron. It contains impurities, especially carbon, silicon, phosphorous and metals from the original ore. To make steel from the pig iron, its carbon content must be reduced to less than 1.4%. Most steel is made by the basic oxygen process, in which the excess carbon in the iron is burned off with oxygen (see Fig. 2.14, p. 813, Chang). The molten pig iron is held in the bottom of a cylindrical vessel. Oxygen is introduced in the vessel through a water-cooled tube, and it reacts with the impurities to form oxides:

$$Si + O_2 = SiO_2 \qquad\qquad (11)$$

$$4P + 5O_2 = 2P_2O_5 \qquad\qquad (12)$$

$$2Mn + O_2 = 2MnO \qquad\qquad (13)$$

$$S + O_2 = SO_2 \qquad\qquad (14)$$

and carbon by reaction 5. The gases (O_2 and CO_2) go off through a stack, and the other oxides combine with CaO to form calcium compounds:

$$P_2O_5 + 3CaO = Ca_3(PO_4)_2 \qquad\qquad (15)$$

or with silica:

$$MnO + SiO_2 = MnSiO_3 \qquad\qquad (16)$$

The resulting molten oxides are withdrawn as slag.

Aluminum is found naturally as bauxite, $Al(OH)_3$. Mild heating decomposes it to alumina and water:

$$2Al(OH)_3 = Al_2O_3 + 3H_2O \qquad\qquad (17)$$

The alumina is reduced in the liquid state to aluminum by electrolysis. The melting point of alumina is very high, 2050°C, which is too high for a commercial process. To reduce the melting point of alumina cryolite, Na_3AlF_6, is added; a mixture of cryolite and alumina melts at a much lower temperature of about 1000°C. Then the alumina is reduced to aluminum by electrolysis:

$$2Al_2O_3 = 4Al + 3O_2 \qquad\qquad (18)$$

The metals made by these various processes are formed into useful shapes by a variety of methods. Examples are rolling to sheet, drawing to wire, molding, forging (stamping or hammering), and extruding through dies. Metals and alloys with high melting points such as tungsten (3410°C), molybdenum (2625°C), tantalum (2996°C) and niobium (2415°C) are often formed as powders, just as for ceramics as described in Section 4.2. The powders are formed into desired shapes and then heated to sinter them together in the solid state at temperatures much below their melting temperatures.

The metals resulting from these processes are pure enough for many applications, and can also serve as components of metallic mixtures. However, for certain applications, it is desirable to have very pure metals. They can be further purified by the following processes:

1. *Distillation.* Metals with low boiling points such as mercury, magnesium, zinc, and lead can be purified by direct vaporization or fractional distillation; the impurities distill off or remain in the solid, and the desired metal is condensed from the vapor. Some metals react with other substances to form low boiling compounds that can be then fractionally distilled. An example is the reaction of nickel with carbon monoxide to form nickel carbonyl at about 70°C:

$$Ni + 4CO = Ni(CO)_4 \tag{19}$$

This compound has a low vapor pressure, and after distillation decomposes to nickel in the reverse of reaction 19 by heating to 200°C.

2. *Electrolysis of molten metal.* This process is described in the chapter on electrochemistry.

3. *Zone refining.* In this process a rod of metal is heated to the melting point at one end, and the molten region then moved through the solid rod as shown in Fig. 2.18, p. 816, Chang. The impurities segregate to the end of the rod, leaving behind purer solid. This process is described in more detail in the chapter on phase diagrams.

10.6 ➤ ELECTRONIC MATERIALS

When I was a youngster, I made a crystal radio set. The heart of this set was a detector crystal of natural galena (lead sulfide, PbS). A fine metal wire (cat's whisker) was scratched over the surface of the crystal to find a sensitive spot, and the crystal incorporated into a tuned circuit. With an antenna of about 12 meters length, enough power from a local radio station could be drawn to supply a pair of ear phones, with no batteries. The galena is a semiconductor, and different regions of it had different concentrations of impurities. Junctions between these regions provide electrical devices that can rectify and amplify radio signals.

Semiconductors with regions of different impurity concentrations are the elements of all electronic circuits. To control the impurity levels the semiconductor must be highly pure. The purification of silicon was the technological advance in the chemistry of materials that has led to the explosion of electronic devices developed since 1945.

For highly pure materials, impurity levels are described as parts per million (ppm) or parts per billion (ppb). These units are weight of impurity per weight of matrix - grams per gram. Perhaps a better way to describe impurity of levels is in fraction of atoms.

Sample Problem:
Calculate the fraction of aluminum or boron atoms in silicon if the impurity level is a) one ppm, b) one ppb.

Solution:
Consider one gram of silicon. The number of atoms of silicon is A/28.09 in which 28.09 is the molecular weight of silicon and A is Avogadro's number of $6.02(10)^{23}$ atoms/mole. The number of atoms of aluminum in one gram of silicon, if the aluminum concentration is one ppm, is $10^{-6}A/26.98$, in which 26.98 is the atomic weight of aluminum. Thus, the atom fraction of aluminum is $10^{-6}(28.09)/26.98 = 1.04(10)^{-6}$. The number is close to the concentration in grams/gram, because the molecular weights of silicon and aluminum are nearly the same. For one ppb, the atom ratio of aluminum to silicon is $1.04(10)^{-9}$. The fraction of boron atoms in silicon when boron is at one ppm is $10^{-6}(28.09)/10.81 = 2.60(10)^{-6}$, where 10.81 is the molecular weight of boron. At one ppb, the atom ratio is $2.60(10)^{-9}$.

To make satisfactory electronic devices the concentrations of most impurities in silicon must be below the ppb range. Achieving this extremely high purity was the prerequisite to the electronics age; the processing of silicon is now described.

The starting material for making silicon is quartz sand. A wag has said that everyone has a fortune in silicon in their backyard - the problem is that it is not pure. Some natural sources of quartz provide one of the purest natural materials known, with impurities in the ppm range; the main impurities are aluminum and sodium. The quartz sand is reacted with carbon (coke) to reduce it to silicon:

$$SiO_2 + 2C = Si + 2CO \tag{20}$$

The impurities in this silicon come both from the quartz and mainly from the coke. In the next step the silicon is purified by reacting it to form a compound with a low boiling point, which can then be fractionally distilled as described above. Two different compounds are possible; one is silicon tetrachloride:

$$Si + 2Cl_2 = SiCl_4 \tag{21}$$

After fractional distillation of the $SiCl_4$, it is reduced with magnesium to silicon:

$$SiCl_4 + 2Mg = Si + MgCl_2 \qquad (22)$$

The magnesium chloride is removed by dissolving it in water.

Another compound is trichlorosilane, $SiCl_3H$:

$$Si + 3HCl = SiCl_3H + H_2 \qquad (23)$$

To recover the silicon from the fractionally distilled $SiCl_3H$, this gas is heated to 1000°C, in hydrogen, which reverses reaction 23 to form silicon. The silicon formed from these gases is called electronic grade, and is pure enough for some electronic devices. For higher purity the silicon is purified by zone melting, as described above and in Chapter ___.

The silicon from these processes is polycrystalline, with random orientation of small grains. However, for controlled electronic properties in devices, the silicon must be all of one crystalline orientation; that is, it must be a <u>single crystal</u>. To grow a single crystal, a small seed single crystal of silicon is touched to the top of a silicon melt, and a single crystal is "pulled" from the melt. This "Czochralski" method can grow cylindrical single crystals of silicon from 10 to 20 cm in diameter. "Chips" of thin silicon layers are then cut from these cylinders. Usually a very small amount of impurity is added to the melt, so that the silicon single crystal has a controlled amount of impurity that gives either positive or negative carriers of electricity in the silicon.

Electronic devices depend on junctions between regions which contain different impurities. In the early days of making devices a region of different impurity concentration was made by adding a different impurity to the melt during growth of the silicon single crystal. This melt growth method was used to make small transistors. As smaller devices become desirable, a region on the surface of a single crystal of silicon was alloyed with a low-melting impurity such as indium. A small piece of the indium was placed on the silicon single crystal at the place where a device was required, and the indium heated to melt it and a portion of the silicon. More control is possible if the impurity is diffused into the silicon from the vapor. For example, volatile compounds such as BBr_3 or BCl_3 for boron impurities are heated at the silicon surface, forming elemental boron on the surface:

$$2BCl_3 = 2B + 3Cl_2 \qquad (24)$$

The boron on the silicon surface diffuses into it, forming a region of higher boron concentration.

A tremendous advance in the number density of devices on a silicon surface has taken place in the last few decades. These high densities are made possible by careful control of impurity depositions on the silicon surface. The details of making these integrated circuits will be described in a later chapter.

TABLE 10-1
Examples of Solid Materials Important in Modern Technology

Ceramics
  Alumina, Al_2O_3
  Brick
  Barium Titanate, $BaTiO_3$
  Cement, mainly calcium
    silicate
  Ferrites (e.g., $NiFe_2O_8$)
  Magnesia, MgO
  Pottery
  Porcelain
  Refractories
    (e.g., $2Al_2O_3$-$3SiO_2$,
    mullite)
  Silicon Carbide, SiC
  Silicon Nitride, Si_3N_4
  Uranium Dioxide, UO_2
  Tile
  Traditional ceramics
    such as brick, pottery,
    porcelain, and tile are
    mainly aluminosilicates.

Electronic Materials
  Silicon
  Germanium
  Gallium Arsenide, GaAs

Glasses
  Soda-lime (75% SiO_2,
    15% Na_2O, 10% CaO)
  Fused silica, SiO_2
  Pyrex borosilicate
  Lead glasses

Metals and Alloys
  Aluminum
  Copper and Brass
  Gold
  Iron and Steel
  Lead
  Nickel
  Silver
  Tin
  Tungsten

Polymers
  Polyethylene
  Polystyrene
  Polyvinyl Chloride
  Rubber
  Nylon
  Cellophane

Composites
  Concrete (sand and rock
    in cement)
  Fiber Glass (glass fiber
    in a polymer)
  Glass ceramics (uniform
    crystals in glass)
  Ceramic Fibers and
    Particles in Metals
  Coatings

TABLE 10-2
Some Important Uses of Solid Materials

Ceramics
- Artware
- Building Materials
- Containers
- Electrical Insulators
- High Temperature
  - Materials
- Lasers
- Magnets
- Optical Elements

Electronic Materials
- Amplifiers
- Rectifiers
- Integrated Circuits
- Switches
- Solar Generators
- Phosphors

Glasses
- Containers
- Coatings
- Crucibles for Melting
  - Semiconductors
- Lamps
- Optical Elements
- Radioactive Waste
  - Disposal
- Windows

Metals and Alloys
- Airplanes
- Automobiles and
  - Trucks
- Building Construction
- Bridges
- Coins
- Engines
- Household Appliances
- Jewelry
- Ships

Polymers
- Boats
- Books
- Coatings
- Electrical Insulation
- Musical Instruments
- Toys
- Furniture
- Packaging

TABLE 10-3
Ceramic and Glass Raw Materials

Name	*Chemical Formula*
Clay Minerals (e.g., Kaolin)	$Al_2O_3 - 2\ SiO_2 - 2H_2O$
Silica Sand	SiO_2
Feldspar, e.g.	$K_2O - Al_2O_3 - 6SiO_2$
Sillimanite	$Al_2O_3 - SiO_2$
Limestone	$CaCO_3$
Dolomite	$CaMg(CO_3)_2$
Soda Ash	Na_2CO_3
Borax	$Na_2B_4O_7 - 10H_2O$
Bauxite	$Al(OH)_3$

TABLE 10-4
Methods for Reducing Metallic Compounds to Metals
(the metals are listed from less to more reactive)

Metals	Compounds	Methods
Gold, silver, copper, mercury, platinum	uncombined	
Copper, silver, lead, mercury, zinc, nickel	Sulfides, e.g., Cu_2S	Heat in air to form oxides, then reduce with carbon, or reduce directly in air
Iron	Oxide, Fe_2O_3	Heat with coke (carbon)
Aluminum	Oxide, Al_2O_3	Reduce electrolytically in the melt
Chromium, manganese, vanadium	Oxide, e.g., Cr_2O_3	Reduce with more reactive metal such as aluminum
Lithium, sodium, magnesium, calcium	Chloride, e.g., NaCl	Reduce electrolytically in melt or solution

11

CHEMICAL KINETICS

11.1 ➤ THERMODYNAMICS VS. KINETICS

Thermodynamics tells us what processes are possible and which are not. Kinetics tells us if a possible process is probable (in other words, will it occur at a reasonable rate?). Kinetic studies of reactions also can reveal something about the mechanism of the reaction, which is important if one ultimately wishes to control the course of a reaction. Thermodynamics tells us nothing about mechanism, only the relative stabilities of final and initial states.

11.2 ➤ FACTORS INFLUENCING REACTION RATE

Consider the following hypothetical reaction occurring in the gas phase:

$$2A \rightarrow B \qquad \qquad (1)$$

We of course must assume that the ΔG is negative, otherwise the reaction is impossible. What factors might influence the rate at which A reacts to form product B? One is the collision frequency. Molecules of A need to 'hit' each other if they are to react. The collision frequency, v, should depend upon the concentration of reactants. Therefore,

$$v \; \alpha \; [A]^2 \qquad \qquad (2)$$

The kinetic theory of gases tells us (Ch. 5 of Chang) that the average velocity, v, of gaseous molecules increases with the square-root of temperature:

$$v \; \alpha \; (T)^{1/2} \qquad \qquad (3)$$

The collision frequency increases with increasing average velocity, and therefore v might also be expected to increase with the square-root of T.

A second point to be considered is that not all collisions will necessarily be productive (meaning not all will necessarily lead to product molecules). The reacting molecules may need to approach each other with the right orientation to react. For example, in the reaction of two H-I molecules to form H_2 and I_2, side-by-side encounters might be expected to be more 'productive' than end-to-end encounters.

However, even a collision that has the right orientation does not guarantee product formation. Why not? Most reactions have an <u>activation energy</u> associated with them that must be overcome before a measurable rate is realized. This means that a collision with the right orientation also must have a certain minimum energy associated with it to be productive. Bonds need to be distorted and partially broken prior to reaction, and this requires energy, even if the reaction itself is exothermic.

The activation energy, E_{act}, can be represented as a barrier which needs to be surmounted as reactants are transformed to products. This is shown in the sketch below, where energy is plotted as a function of 'progress of reaction,' the latter axis representing the conversion of reactants to products.

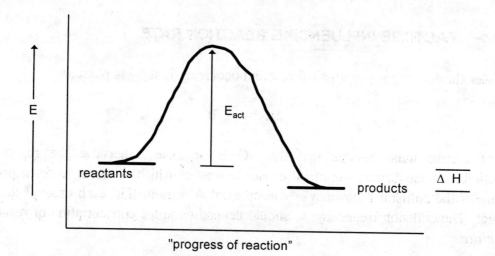

As a general rule of thumb for many reactions, a 10 K rise in temperature increases the rate by 2-4 times. We will show in lecture that this large increase can't be the result of an increase in collision frequency. (Try proving it using Equation 3.) The fraction of collisions with the appropriate energy ($E \geq E_{act}$) increases with temperature also, and this is key reason why the rate increases with T. This is true even for exothermic reactions, although here an increase in temperature shifts the reaction to the left. In this case, products form more quickly but the yield decreases with increasing temperature, and choices have to be made about the relative merits of forming less product faster.

We can summarize all of the previous discussion as follows:

$$\text{Rate} \ \alpha \ (\text{\# of collisions}) \times (\text{'orientation factor'}) \times (\text{\# of molecules with } E > E_{act}) \quad \textbf{(4)}$$

11.3 ➤ DETERMINATION OF REACTION RATES— SPECTROPHOTOMETRY

The rate of a reaction can be obtained by monitoring the change in concentration of a reactant or a product with time. There are many ways to do this, but probably the most convenient and widely used is spectrophotometry. Molecules have unique absorptions in various regions of the electromagnetic spectrum, and these can be used as fingerprints to follow concentration changes. For example, in the uv-visible range, electrons can be promoted to higher energy states (absorption), and fall back to the ground state (emission), and these events occur with characteristic frequencies. (Absorption is more commonly used in the laboratory.) Also, molecules whose bond vibrations are accompanied by a change in dipole moment absorb IR radiation, also with characteristic frequencies. An instrument known as a spectrophotometer is capable of generating a broad range of frequencies (or wavelengths) and a detector monitors the intensity of light as a function of frequency or wavelength. The ratio of the intensity of the transmitted light (I) to the incident intensity (I_0) is the transmittance (T).

$$T \ = \ I/I_0 \qquad\qquad \textbf{(5)}$$

The log of the inverse of the transmittance is the absorbance (A), and it is related to concentration through Beer's law:

$$\log(I_0/I) \ = \ A \ = \ \varepsilon \, c \, l \qquad\qquad \textbf{(6)}$$

Here c is the concentration of the species giving rise to the absorption, l is the path length of the cell holding the sample (say 1 cm or 0.1 cm), and ε is the molar extinction coefficient. The latter is related to the probability of a given electronic or vibrational transition, and is different for each bond giving rise to a particular absorption. If ε and l are known, the concentration of a particular species can be tracked by monitoring the change in absorbance with time. A typical plot for the disappearance of species A in Eq. 1 can be found in Chang, Fig. 13-2.

11.4 ➤ REACTION RATES AND REACTION ORDERS

Referring to the preceding figure, we will define the instantaneous reaction rate as the limiting value of the change in [A] with time, i.e. $-d[A]/dt$. (We could equally well use the increase in product concentration, [B], to define the rate. For hypothetical reaction 1 (Eq. 1), it would be $d[B]/dt$.) The instantaneous rate is expected to depend on concentration. A couple of possibilities are

$$-d[A]/dt = k\,[A] \tag{7}$$

or

$$-d[A]/dt = k\,[A]^2 \tag{8}$$

Equations 7 and 8 are examples of a <u>rate law</u>, which governs the dependence of rate on concentration (reactant in this case.) Since the rate in Eq. 7 is proportional to the first power of [A], the rate law is said to be first-order in [A]. In Eq. 8 the rate is proportional to the second power of [A], the rate law is said to be second-order in [A]. Notice that the ratio $(-d[A]/dt)/[A]$ in Eq. 7 and $(-d[A]/dt)/[A]^2$ in Eq. 8 is a constant, k (though not necessarily the same constant in both cases). This is the <u>rate constant</u>. It has the following temperature dependence:

$$k = A\exp\,(-E_a/RT) \tag{9}$$

where A is the pre-exponential factor (essentially a composite of the collision frequency and 'orientation factor' in Eq. 4), E_a is the activation energy, R is the gas constant (8.314 J/mol-K), and T is temperature in K). The term $\exp\,(-E_a/RT)$ is proportional to the fraction of the total number of collisions having $E \geq E_{act}$. Eq. 9 is frequently referred to as the <u>Arrhenius equation</u>.

If the reaction occurs in a single step (a so-called elementary reaction), then the exponents in the rate law can be derived from the coefficients of the reaction. Therefore, for the reaction in Eq. 1, the rate law in Eq. 8 would be the correct one. In the reaction $2\,NOCl \rightarrow 2\,NO + Cl_2$, the rate law appears like it might be

$$-d[NOCl]/dt = k\,[NOCl]^2 \tag{10}$$

and this is what is found experimentally. The rate law is said to be second-order in [NOCl]. However, for reactions which do not proceed in one elementary reaction, the rate law must be determined experimentally.

One way to determine the reaction order is as follows. (Let's focus on the NOCl reaction mentioned above.) The proposed second-order rate law can be integrated to give

$$1/[NOCl]_t \; = \; 1/[NOCl]_0 \; + \; kt \qquad\qquad (11)$$

where $[NOCl]_t$ and $[NOCl]_0$ are the concentrations at some time t and at t = 0, respectively. A plot of $1/[NOCl]_t$ vs. t should be linear is the reaction is truly second-order and the rate law in Eq. 10 is correct. In addition, such a plot would give the rate constant, k, from the slope. This is in fact how rate constants are commonly determined.

11.5 ➤ AN APPLICATION OF A FIRST-ORDER RATE LAW: RADIOCARBON DATING

There are three known isotopes of carbon, ^{12}C, ^{13}C, and ^{14}C. The last is radioactive, and the determination of its amount in dead organic matter affords the possibility of estimating the age of the matter.

^{14}C is constantly being generated in the atmosphere from the reaction of ^{14}N with neutrons. ^{14}C is radioactive, indicating it is not stable. In fact, it decays back to ^{14}N with the emission of a β particle (an electron). The net result over a long period of time is that the $[^{14}C]$ in the atmosphere is ≈ constant. Some of this radioactive carbon reacts with O_2 to form radioactive CO_2, or $^{14}CO_2$, which is incorporated into plants through photosynthesis and which is eventually eaten by us. Some radioactive CO_2 is released back into the atmosphere by us during respiration. The net result is that <u>in living species</u> the $^{14}C/^{12}C$ ratio is ≈ constant.

When a species dies, respiration no longer occurs, and therefore the ^{14}C concentration in the body slowly decreases as the ^{14}C disintegrates to ^{14}N. Because radioisotope decay follows first-order kinetics, it is possible to estimate the age of dead organic matter by monitoring the ratio of ^{14}C to ^{12}C and comparing it to the typical ratio found in living matter.

The rate law for the decay of any isotope I is

$$-d[I]/dt = k [I] \qquad (12)$$

Separating variables followed by integration gives

$$\ln [I]_t = -kt + \ln [I]_o \qquad (13)$$

Rearrangement gives

$$\ln \{[I]_o/[I]_t\} = kt \qquad (14)$$

When $[I]_t = 1/2 [I]_o$, $\ln \{[I]_o/[I]_t\} = \ln 2$. Here t is defined as the half-life, and is designated as $t_{1/2}$.

$$\ln 2 = kt_{1/2} \qquad (15)$$

The half-life for ^{14}C is known to be 5730 yr. Using Eq. 4, the first-order rate constant k is determined to be 1.21×10^{-4} yr^{-1}.

11.6 ➤ CATALYSIS

Many chemical reactions of industrial significance are catalyzed - that is, they utilize catalysts to accelerate their rates. A good definition of a catalyst is the following: <u>it is a substance that increases the rate of reaction without being consumed</u>. Catalysts only increase reaction rate—they have no effect on the position of the equilibrium and therefore do not appear in equilibrium constant expressions. It follows that they have no effect on the ΔG of a reaction. Catalysts may be of the same phase as the reactants (homogeneous catalysts) or a different phase, usually a solid (heterogeneous catalysts). Examples of the latter include finely divided Pt metal supported on aluminum oxide which is used in 'catalytic converters' in automobiles.

We mentioned during the last lecture that <u>activation energy</u> plays the dominant role in determining reaction rates. Recall that bonds in the reactants frequently need to be partially broken or strained prior to reaction—this is the typical origin of activation energy. Catalysts provide an alternative pathway for the reaction, one in which the bonds in the reactants are 'loosened' or 'activated,' making the reaction occur more quickly. The activation energy of this alternate pathway is lower (see figure below), and hence the reaction is faster.

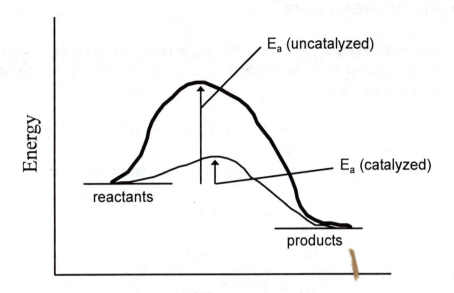

Note that a catalyst for the forward reaction is also a catalyst for the reverse reaction.

Enzymes are the best catalysts known. They are large molecules which increase reaction rates by many powers of ten, and many are also selective—they 'work' on only one type of reactant (called a substrate) out of many with rather similar structures.

A simplistic picture of enzyme action is shown in the sketch below. The enzyme functions as a template, bringing the reactants together in close proximity. The binding sites in the enzyme are complementary in shape to the reactants and this is what accounts for the selectivity of enzymes (Chang, Fig. 13.22). The enzyme-substrate complex is bound in such a way that reactant bonds are activated towards product formation.

How effective are enzymes at increasing reaction rates? Consider the following reaction:

$$CO_2 \ + \ H_2O \ \rightarrow \ H_2CO_3 \tag{16}$$

This reaction is of great importance in quickly moving carbon dioxide from sites where it is formed in tissues to the lungs to be expelled. This reaction is catalyzed in our bodies by an enzyme (carbonic anhydrase), which accelerates the rate by 1×10^7 vs. the uncatalyzed reaction. About 100,000 molecules of reactant can be converted to product in 1 second (per enzyme molecule).

11.7 ➤ CHAIN REACTIONS

Consider the chlorination of methane to give methyl chloride. It occurs upon irradiation with ultraviolet light at room temperature or at high temperatures in the absence of light. The overall reaction is

$$CH_4 + Cl_2 \rightarrow CH_3Cl + HCl \tag{17}$$

Kinetic studies have revealed the following mechanism for the reaction.

$$Cl_2 \rightarrow 2\ Cl\cdot \tag{18}$$

$$CH_4 + Cl\cdot \rightarrow \cdot CH_3 + HCl \tag{19}$$

$$\cdot CH_3 + Cl_2 \rightarrow CH_3Cl + Cl\cdot \tag{20}$$

$$\cdot CH_3 + Cl\cdot \rightarrow CH_3Cl \tag{21}$$

$$2\ Cl\cdot \rightarrow Cl_2 \tag{22}$$

$$2\cdot CH_3 \rightarrow C_2H_6 \tag{23}$$

The first step, cleavage of a Cl-Cl bond, is the <u>initiation</u> step (Eq. 18). Cl atoms ($\cdot$Cl) then abstract a hydrogen atom from methane to form the methyl radical ($\cdot CH_3$) and HCl (Eq. 19). The methyl radical reacts with chlorine to produce methyl chloride plus chlorine atoms (Eq. 20). A key point is that $\cdot$Cl is regenerated and can react again as shown in Eq. 19. Notice that if Eqs. 19 and 20 keep going, many molecules of product (up to about 10,000) can be formed per Cl atom. This is an example of a <u>chain reaction</u>. Equations 19 and 20 are called <u>propagation</u> steps. Eventually, the reaction suffers termination, which here involves consumption of the reactive species $\cdot$Cl and $\cdot CH_3$ (Eqs. 21-23). <u>All chain reactions have three steps: initiation, propagation, and termination</u>. Chain reactions are of great industrial importance, particularly in the synthesis of polymers such as polystyrene (Styrofoam), polyethylene (GladWrap, Baggies), and polyvinyl chloride (PVC). We will learn about polymerization chemistry next semester.

11.8 ➤ KINETICS SUMMARY

- Reaction rate depends on collision frequency, orientation factor and the fraction of molecules or atoms with $E \geq E_a$. The latter is typically most important. Have a mental picture of the meaning of activation energy.

- The fraction of molecules or atoms with $E \geq E_a$ increases <u>exponentially</u> with temperature.

- The rate of a reaction is written as -d[reactant]/dt or d[product]/dt.

- A rate law shows the dependence of rate on concentration of products or reactants. The exponents of the concentration terms are the same as the coefficients of the reactants or products only if the reaction is elementary (i.e., it proceeds in a single step). Otherwise the exponents must be determined by experiment. (See the HW problem on the dehydration of grain alcohol.)

- Two key rate laws (k is the rate constant):

    First-order reaction: $-d[A]/dt = k[A]$; k has units of sec^{-1}
    Second-order reaction: $-d[A]/dt = k[A]^2$; k has units of l/mol-sec.

- The rate constant increases with temperature as

$$k = A \exp(-E_a/RT)$$

- The integrated form of a rate law can be used to determine the order of a reaction. Also, the rate constant can be determined once the order is established (e.g., by plotting $\ln[A]_t$ vs. t if first-order, for example. Slope is -k).

- For a first-order reaction, $t_{1/2} = 0.693/k$, where $t_{1/2}$ is the half-life.

- A catalyst affords an alternate, lower activation energy pathway for a reaction. It does not change the position of the equilibrium.

- Chain reactions have three steps: initiation, propagation, and termination.